Mathematics Study Resources

Volume 8

Series Editors

Kolja Knauer, Departament de Matemàtiques Informàtic, Universitat de Barcelona, Barcelona, Barcelona, Spain

Elijah Liflyand, Department of Mathematics, Bar-Ilan University, Ramat-Gan, Israel

W0230021

This series comprises direct translations of successful foreign language titles, especially from the German language.

Powered by advances in automated translation, these books draw on global teaching excellence to provide students and lecturers with diverse materials for teaching and study.

Oliver Stein

Basic Concepts of
Nonlinear Optimization

 Springer

Oliver Stein (ID)
Institut für Operations Research (IOR)
Karlsruher Institut für Technologie (KIT)
Karlsruhe, Germany

ISSN 2731-3824   ISSN 2731-3832   (electronic)
Mathematics Study Resources
ISBN 978-3-662-69740-5   ISBN 978-3-662-69741-2   (eBook)
https://doi.org/10.1007/978-3-662-69741-2

This book is a translation of the original German edition "Grundzüge der Nichtlinearen Optimierung," 2nd edition, by Oliver Stein, published by Springer-Verlag GmbH, DE in 2021. The translation was done with the help of an artificial intelligence machine translation tool. A subsequent human revision was done primarily in terms of content, so that the book will read stylistically differently from a conventional translation. Springer Nature works continuously to further the development of tools for the production of books and on the related technologies to support the authors.

Translation from the German language edition: "Grundzüge der Nichtlinearen Optimierung" by Oliver Stein, © Springer-Verlag GmbH Deutschland, ein Teil von Springer Nature 2021. Published by Springer-Verlag GmbH. All Rights Reserved.

This Springer imprint is published by the registered company Springer-Verlag GmbH, DE, part of Springer Nature.
The registered company address is: Heidelberger Platz 3, 14197 Berlin, Germany

If disposing of this product, please recycle the paper.

Here we are now, entertain us.
(Marc-Uwe Kling)

Preface

This textbook originated from the notes of my lecture 'Nonlinear Optimization I and II', which I have been giving annually at the Karlsruhe Institute of Technology since 2006. The primary audience for this lecture are students of Industrial Engineering and Management in the Bachelor's specialization program. In this textbook, this is reflected in that mathematical facts are treated stringently, but are significantly more motivated and illustrated than in a textbook for a purely mathematical course of study. Therefore, this book is aimed at students who want to understand and apply mathematically founded methods in their course of study, as is the case in the natural, engineering, and economic sciences. Since the more detailed motivation naturally comes at the expense of the scope of the material, this book limits itself to the presentation of the *basic concepts* of nonlinear optimization.

The subject is the treatment of minimization or maximization models with nonlinear objective functions under nonlinear constraints, as they often occur in application disciplines. For such problems, we derive optimality conditions and provide algorithms based on them.

A common feature of all these optimality conditions and algorithms is that they are based on the evaluation of first or second derivatives of the functions involved. While this often makes the optimality conditions easy to handle, they suffer from the fundamental problem that derivatives only reflect the shape of functions locally, not globally. Accordingly, the solution methods based on these optimality conditions are only able to identify *local* optimal points. Only in exceptional cases are these also *global* optimal points. A particular such exception is given by convex optimization problems, where every local minimal point is automatically also a global minimal point.

In the present text we take the view that in applications the knowledge of local optimal points is often valuable, especially if they can be calculated quickly. Nevertheless, often also the algorithmic determination of global optimal points is possible, but the effort required is usually much higher than that for determining local optimal points. Algorithms for determining global minimal points of nonconvex problems are discussed in detail in [36]. They partly involve the solution of auxiliary problems by the local methods presented here.

Before dealing with the theoretical and algorithmic identification of local optimal points, however, it is important to clarify whether an optimization problem possesses optimal points in the first place. The introductory Chap. 1 therefore covers some sufficient conditions for solvability, in addition to basic terminology and notation.

Chapter 2 focuses on nonlinear optimization problems without constraints and derives optimality conditions for them, based on first and second derivatives of the objective function. The simplifications that arise in the case of minimizing a convex objective function are briefly mentioned. Based on the optimality conditions, we then formulate and discuss a number of important algorithms—from the gradient method dating back to the mid-nineteenth century to modern trust-region methods.

Chapter 3 extends the considered optimization models to include nonlinear constraints. However, deriving meaningful optimality conditions here requires considerably more effort than in the unconstrained case from Chap. 2. This is due to the fact that the location of optimal points depends only on the *geometry* of the set of feasible points, but different *functional descriptions* of this set by constraints can lead to unpleasant effects in the respective optimality conditions. Chapter 3 discusses this issue in detail and derives the corresponding optimality conditions before we again discuss various algorithms based on them.

This textbook can serve as the basis for a four-hour lecture. It partly relies on presentations by the authors W. Alt [1], M.S. Bazaraa, H.D. Sherali and C.M. Shetty [2], A. Beck [3], U. Faigle, W. Kern and G. Still [7], O. Güler [16], H.Th. Jongen, K. Meer and E. Triesch [23] as well as J. Nocedal and S. Wright [26], who also deal with many questions going beyond this book. For basics of convex and global optimization, as mentioned, we refer to [36], and for general basics of optimization to [25].

The present textbook does not aim at giving a comprehensive literature review. In particular, since it is based on an automated translation of the German original textbook version, several other German books are cited, partly authored by myself. For the latter, please check for translations generated after the publication of the present book.

At this point, I would like to thank the Springer staff for their very helpful cooperation in translating the manuscript and copy editing.

A big thank you also goes to my current and former PhD students Dr. Tomáš Bajbar, Prof. Dr. Peter Kirst, Dr. Robert Mohr, Dr. Christoph Neumann, Stefan Schwarze, Dr. Marcel Sinske, Dr. Paul Steuermann and Prof. Dr. Nathan Sudermann-Merx as well as numerous students who have pointed out to me possibilities for content and formal improvements during the development of this teaching material. The present text was typeset in LaTeX2e. The illustrations come from *Xfig*.

Text set in smaller typeface indicates material that is given for completeness but can be skipped on first reading.

Karlsruhe, Germany   Oliver Stein
May 2024

Contents

Introduction

1

Contents

Finite-dimensional continuous optimization deals with the minimization or maximization of an objective function in a finite number of continuous decision variables. Important applications can be found not only in linear models (as in simple models for profit maximization in production programs or in transportation problems [25]), but also in various nonlinear models from natural, engineering, and economic sciences. These include geometric problems, mechanical problems, parameter-fitting problems, estimation problems, approximation problems, data classification, and sensitivity analysis. It is also used as a solution tool in noncooperative games [34], in robust optimization [34], or in the continuous relaxation of discrete and mixed-integer optimization problems [35].

This introductory chapter first motivates in Sect. 1.1 the basic terminology and notation of optimization problems using various examples, and distinguishes finite-dimensional smooth optimization from infinite-dimensional and nonsmooth optimization. Section 1.2 then addresses the question under what conditions optimization problems can be solved at all (a much more detailed presentation of solvability issues can be found in [36]). Finally, Sect. 1.3 provides some rules of calculus and transformations for optimization problems, which play a role in the context of this textbook.

1.1 Examples and Terminology

In optimization, we compare different alternatives with respect to an objective criterion and search for a best among all considered alternatives. Along the following example of a nonlinear optimization problem in two variables, which can be solved using school-level mathematics, we introduce some basic terminology.

O. Stein, *Basic Concepts of Nonlinear Optimization*,
Mathematics Study Resources 8, https://doi.org/10.1007/978-3-662-69741-2_1

Example 1.1.1 (Tin Can—Smooth Optimization) From A units (e.g., square centimeters) of sheet metal, a tin can with maximal volume is to be constructed. The can is modeled as a cylinder with lid and bottom, so that it is completely characterized by two specifications, namely its radius r and its height h. The can then has the volume $V(r, h) = \pi r^2 h$ and the surface area $2\pi rh + 2\pi r^2$.

The objective criterion to be maximized here is thus the volume V of the can, and the feasible alternatives are the pairs $(r, h) \in \mathbb{R}^2$ in the set described by the constraints

$$M = \{(r, h) \in \mathbb{R}^2 \mid 2\pi rh + 2\pi r^2 \leq A, \ r, h \geq 0\}.$$

The structurally simple constraints $r, h \geq 0$, which are frequently present also in other optimization models, are called *nonnegativity constraints*.

As in this example, in continuous optimization problems the alternatives can always be interpreted geometrically as 'points in a space', here in the two-dimensional Euclidean space $\mathbb{R}^2$. Since exactly this geometric correspondence will motivate optimality conditions and algorithms, we will no longer address the elements of M as feasible alternatives, but as *feasible points*. We will call the set M the *feasible set*.

In every optimization problem, the objective criterion assigns a scalar value to each feasible point and is therefore, from a mathematical point of view, a function from M to the set $\mathbb{R}$ of real numbers. We call this function the *objective function*. In the present example, this is the volume function

$$V : M \to \mathbb{R}, \ (r, h) \mapsto \pi r^2 h.$$

We write the general task of maximizing a function f over a set M as an optimization problem in the form

$$P : \quad \max \ f(x) \quad \text{s.t.} \quad x \in M.$$

The abbreviation *s.t.* stands for *subject to* (or *so that*) and indicates that in the formulation of P the description of the feasible set starts here. A minimization problem would be written analogously in the form

$$P : \quad \min \ f(x) \quad \text{s.t.} \quad x \in M.$$

We will see that it is actually sufficient to be able to handle only minimization problems.

If an explicit description of M by constraints is given, it is sufficient to specify these constraints in the associated optimization problem P in the place of M. The optimization problem for maximizing the tin can volume is therefore

$$P_{\text{can}} : \quad \max_{r,h} \ \pi r^2 h \quad \text{s.t.} \quad 2\pi rh + 2\pi r^2 \leq A, \quad r, h \geq 0.$$

While in P_{can} a decision is to be made about the values of r and h, the value of A is exogenously given. We therefore call r and h *decision variables* and A a *parameter*. For clarity, the decision variables are often noted below the optimization rule 'max' or 'min', as we did in P_{can}.

A point $\bar{x} \in M$ is called *optimal* for a general optimization problem P, if no point $x \in M$ has a better objective function value. In maximization problems, this means that the inequality $f(x) \leq f(\bar{x})$ is fulfilled for all $x \in M$, and in minimization problems this inequality is reversed. The associated *optimal value* of P is the number $v = f(\bar{x})$. While an optimization problem may possess multiple optimal points, the optimal value is always unique.

We will often express the optimal value of a maximization problem by the notation

$$v = \max_{x \in M} f(x).$$

Here, the 'max' in the above optimization problem P is to be understood as the *task* of maximizing f over M, while the 'max' in the optimal value v denotes a *number*.

To determine an optimal point and the optimal value of the example problem P_{can}, we proceed somewhat laxly for this introductory purpose and use, among other things, calculus concepts from school mathematics, which are only treated later, and in a much more general framework, in Sect. 2.1.

To begin with, one observes that for an optimal choice of decision variables, both $r > 0$ and $h > 0$ will hold. The nonnegativity conditions are then called *inactive*. In contrast, the inequality

$$2\pi rh + 2\pi r^2 \leq A$$

which bounds the size of the can surface should be fulfilled with equality in an optimal point, i.e., be *active*, since excess sheet metal could be used to increase the can volume. The resulting equation can even be explicitly solved for one variable in this case, for example to

$$h = \frac{A}{2\pi r} - r.$$

For these points (r, h) the objective function can therefore be written just as well as

$$V(r, h) = V\left(r, \frac{A}{2\pi r} - r\right) = \frac{A}{2}r - \pi r^3.$$

Therefore, the optimal radius of the initial problem also solves the problem

$$\max_{r} \frac{A}{2}r - \pi r^3 \quad \text{s.t.} \quad r > 0.$$

Fig. 1.1 Tin can volume as a
function of its radius

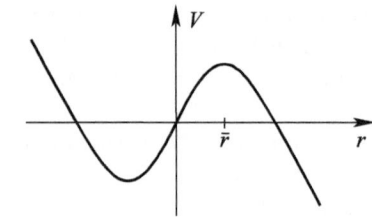

Fig. 1.2 A global and a local
maximal point

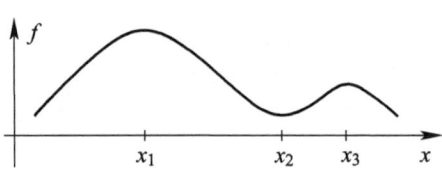

Setting the first derivative to zero for the objective function illustrated in Fig. 1.1, the optimal radius is found to be

$$\bar{r} = \sqrt{\frac{A}{6\pi}},$$

from which

$$\bar{h} = \frac{A}{2\pi\bar{r}} - \bar{r} = \sqrt{\frac{3A}{2\pi}} - \sqrt{\frac{A}{6\pi}} = 2\sqrt{\frac{A}{6\pi}}$$

follows. The *optimal point* of the initial problem is therefore unique and reads

$$\begin{pmatrix} \bar{r} \\ \bar{h} \end{pmatrix} = \sqrt{\frac{A}{6\pi}} \begin{pmatrix} 1 \\ 2 \end{pmatrix},$$

i.e., in particular, the height and diameter of the can are identical in the optimal point. As the *optimal value*, i.e., the maximal can volume achievable with the optimal point, we obtain

$$\max_{(r,h)\in M} V(r, h) = V(\bar{r}, \bar{h}) = \frac{A^{3/2}}{\sqrt{54\pi}}.$$

While the computations for determining the optimal radius in Example 1.1.1 provided the globally optimal radius, other effects often occur when dealing with nonlinear functions, such as the one shown in Fig. 1.2. Here, setting the first derivative of f to zero yields the three solution candidates x_1, x_2, and x_3, where x_2 is not a maximal point due to the positive second derivative $f''(x_2)$. On the other hand, because of $f''(x_1) < 0$ and $f''(x_3) < 0$, both x_1 and x_3 are candidates for maximal

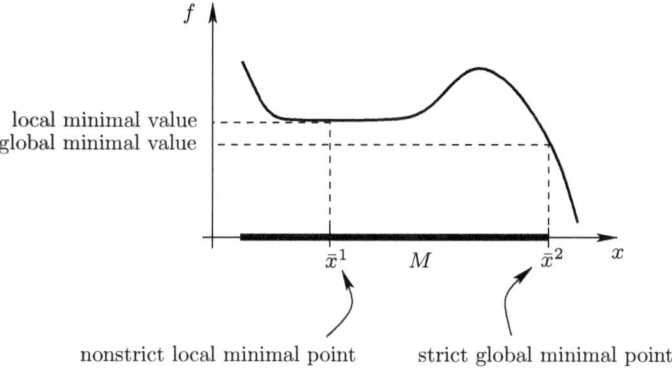

Fig. 1.3 Local and global minimality

points of f. However, x_3 is only the best point among all points in its 'neighborhood' (i.e., a small interval around x_3), while x_1 is optimal among *all* points in $M = \mathbb{R}$.

This important distinction between *local* and *global* optimality is formally established in the following definition (Fig. 1.3). It focuses on minimization problems, while the treatment of maximization problems is discussed subsequently.

Definition 1.1.2 (Minimal Points and Minimal Values) Let a set of feasible points $M \subseteq \mathbb{R}^n$ and an objective function $f : M \to \mathbb{R}$ be given.

(a) $\bar{x} \in M$ is called a *local minimal point* of f on M, if there exists a neighborhood U of $\bar{x}$ with

$$\forall\, x \in U \cap M : \quad f(x) \geq f(\bar{x}).$$

(b) $\bar{x} \in M$ is called a *global minimal point* of f on M, if in part a $U = \mathbb{R}^n$ can be chosen.

(c) A local or global minimal point is called *strict*, if in part a or part b, respectively, for $x \neq \bar{x}$ even the strict inequality $>$ holds.

(d) For each global minimal point $\bar{x}$, $f(\bar{x})$ ($= v = \min_{x \in M} f(x)$) is called *global minimal value*, and for each local minimal point $\bar{x}$, $f(\bar{x})$ is called *local minimal value*.

Regarding the definition of minimal points and values, we note the following:

- For the requirement $f(x) \geq f(\bar{x})$ to make sense, the image space of f must be ordered. For example, the minimization of $f : \mathbb{R}^n \to \mathbb{R}^2$ is not meaningful.

Fig. 1.4 Maximization of f
by minimization of $-f$

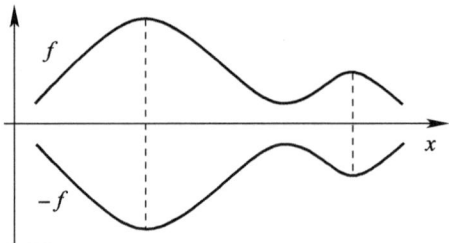

However, the field of *multicriteria optimization* deals with how such problems can still be handled (for a brief introduction see, e.g., [25]).

• Every global minimal point is also a local minimal point.
• Strict global minimal points are unique, and strict local minimal points are locally unique.
• Local and global *maximal* points are defined analogously. Since the maximal points of f are exactly the minimal points of $-f$, it is sufficient to consider minimization problems. Attention: Due to $\max f(x) = -\min(-f(x))$ this construction changes the sign of the optimal value. This is illustrated in Fig. 1.4, proven in Exercise 1.1.3, and more generally in Exercise 1.3.1.
• In view of the similar notation, there is a risk of confusion between the minimal *value* $\min_{x \in M} f(x)$ and the underlying minimization *task P* (cf. the discussion in Example 1.1.1).

Exercise 1.1.3 Given a set of feasible points $M \subseteq \mathbb{R}^n$ and an objective function $f : M \to \mathbb{R}$, show:

(a) The global maximal points of f on M are exactly the global minimal points of $-f$ on M.
(b) If f has global maximal points, the global maximal value is

$$\max_{x \in M} f(x) = -\min_{x \in M} (-f(x)).$$

Algorithms for computing local or global minimal points are not based directly on Definition 1.1.2, as this would require to verify infinitely many inequalities. Instead, we will develop derivative-based optimality conditions and corresponding solution methods in Chaps. 2 and 3. Needless to say, such algorithms are not applicable when the defining functions of the considered optimization problem are not differentiable. The following example illustrates, however, that in some important cases nondifferentiable optimization problems can be transformed into equivalent differentiable problems.

Example 1.1.4 (Discrete Approximation—Smooth vs. Nonsmooth Optimization) Let data points $(x_j, y_j) \in \mathbb{R}^2$, $1 \leq j \leq m$, be given. The goal is to find a straight line that approximates these points 'as well as possible'. If one models the line as the graph of a linear function, i.e., of the form $y = ax + b$, then a pair $(a, b) \in \mathbb{R}^2$ is sought that minimizes the norm of the error vector:

$$\min_{a,b} \left\| \begin{pmatrix} ax_1 + b - y_1 \\ \vdots \\ ax_m + b - y_m \end{pmatrix} \right\|.$$

The choice of the norm is crucial not only for modelling how 'as well as possible' is to be understood, but also for the application of solution methods. For example, for the Euclidean norm $\| \cdot \|_2$ the problem

$$\min_{a,b} \sqrt{\sum_{j=1}^{m} (ax_j + b - y_j)^2}$$

arises. The objective function of this problem is not differentiable everywhere, as the function $\| \cdot \|_2$ is not differentiable at the origin. The above minimization is thus referred to as a *nonsmooth* optimization problem. However, most of the techniques presented in this textbook will refer to *smooth* optimization problems, where at least first derivatives of the involved functions exist.

Fortunately, the above problem can be transformed into an equivalent smooth problem: According to Exercise 1.3.4, the positions of the minimal points are not affected if the root is omitted in the objective function. The new objective function is then differentiable everywhere. With the setting $r_j(a, b) := ax_j + b - y_j$, $1 \leq j \leq m$, it has the structure $\sum_{j=1}^{m} r_j^2(a, b)$, and one speaks of a *least squares problem*.

For the Chebyshev norm $\| \cdot \|_\infty$, one obtains the nonsmooth problem

$$\min_{a,b} \max_{1 \leq j \leq m} |ax_j + b - y_j|,$$

to which one can formulate an equivalent smooth problem by another 'trick', namely by the so-called epigraph reformulation (Exercise 1.3.5). It first provides the equivalent problem

$$\min_{a,b,c} c \quad \text{s.t.} \quad \max_{1 \leq j \leq m} |ax_j + b - y_j| \leq c,$$

whose nonsmooth constraint can then be reformulated to

$$|ax_j + b - y_j| \leq c, \ 1 \leq j \leq m,$$

and, in a final step, to

$$ax_j + b - y_j \le c, \ 1 \le j \le m,$$
$$-(ax_j + b - y_j) \le c, \ 1 \le j \le m.$$

So, one obtains a linear optimization problem with $2m$ constraints in total.

The fact that finite-dimensional optimization problems can also have 'infinite aspects' is demonstrated by the next example. There, in contrast to Example 1.1.1, not finitely many (namely three), but infinitely many inequality constraints occur in a natural manner.

There and in the following we employ a linguistic construction for negations, common in the mathematical literature. For example, we use the artificial term 'nonempty' instead of 'not empty', so that it is clear which word the 'not' refers to.

Example 1.1.5 (Continuous Approximation—Semi-infinite Optimization) Let a nonempty and compact set $Z \subseteq \mathbb{R}^m$, a smooth function $f : Z \to \mathbb{R}$ as well as a family of smooth functions $a(p, \cdot)$ with family parameter $p \in P \subseteq \mathbb{R}^n$ be given (e.g. for $m = 1$ polynomials $a(p, z) = \sum_{j=0}^{n-1} p_j z^j$ of maximal degree $n - 1$). The best approximation to f on Z by a function $a(p, \cdot)$ in the Chebyshev norm is sought. A formulation as an optimization problem is

$$\min_{p \in \mathbb{R}^n} \underbrace{\|a(p, \cdot) - f(\cdot)\|_{\infty,Z}}_{:= \max_{z \in Z} |a(p,z) - f(z)|} \quad \text{s.t.} \quad p \in P.$$

By epigraph reformulation we obtain the equivalent optimization problem

$$\min_{(p,q) \in \mathbb{R}^n \times \mathbb{R}} q \quad \text{s.t.} \quad p \in P, \quad \pm(a(p, z) - f(z)) \le q \quad \forall z \in Z.$$

In this problem, an *infinite* number of inequality constraints is imposed on a *finite*-dimensional decision variable. Such problems are called *semi-infinite* (for details see, e.g., [31]).

The final example of this section illustrates the case of infinite-dimensional optimization, which this textbook does not cover, although some techniques from the finite-dimensional case can be transferred to it (for details see, e.g., [19]). The following optimization problem posed by Johann Bernoulli in 1696 is considered a significant starting point for the development of calculus.

Example 1.1.6 (Brachistochrone—Calculus of Variations, Infinite Optimization) Let two points A and B be given in a vertical plane, with B laterally below A. The task is to find a curve through A and B, such that a mass point moving along

Fig. 1.5 Problem of the brachistochrone

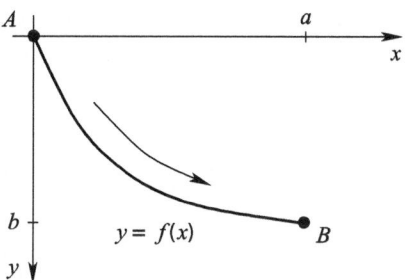

this curve under the influence of gravity covers the distance from A to B in the shortest possible time.

If we place A as in Fig. 1.5 at the origin of the coordinate system, describe the curve as the graph of a function by $y = f(x)$, and assign the coordinates (a, b) to point B with the y-axis pointing downwards, then physical laws yield the value

$$\int_0^a \frac{\sqrt{1 + (f'(x))^2}}{\sqrt{2gf(x)}}\, dx$$

for the transit time of the curve, where g denotes the gravitational constant. The corresponding minimization problem is therefore

$$\min_f \int_0^a \frac{\sqrt{1 + (f'(x))^2}}{\sqrt{2g\, f(x)}}\, dx \quad \text{s.t.} \quad f(0) = 0, \ f(a) = b.$$

The decision variable f stems from the space of differentiable functions, which is an infinite-dimensional space. Therefore, we speak of *infinite-dimensional* optimization.

1.2   Solvability

Whether an optimization problem possesses optimal points at all is not always obvious and, in many solution methods, must be checked in advance by the user. A detailed discussion of this issue can be found in [36], while here we only touch on the main points.

Without any assumptions on the set $M \subseteq \mathbb{R}^n$ and the function $f : M \rightarrow \mathbb{R}$, we can assign a 'generalized minimal value' to every minimization problem

$$P: \quad \min f(x) \quad \text{s.t.} \quad x \in M$$

namely the *infimum* of f on M. To formally introduce it, we call $\alpha \in \mathbb{R}$ a *lower bound* for f on M, if

$$\forall\, x \in M : \quad \alpha \le f(x)$$

applies. The infimum of f on M is the *greatest* lower bound of f on M. Therefore $v = \inf_{x \in M} f(x)$ holds, if

- $v \le f(x)$ for all $x \in M$ (i.e., v is itself a lower bound of f on M) and
- $\alpha \le v$ for all lower bounds α of f on M.

Analogously, the *supremum* $\sup_{x \in M} f(x)$ of f on M is defined as the *smallest upper bound*.

Example 1.2.1 We have $\inf_{x \in \mathbb{R}} (x-5)^2 = 0$ and $\inf_{x \in \mathbb{R}} e^x = 0$.

If f is not bounded from below on M, one formally sets

$$\inf_{x \in M} f(x) \;=\; -\infty,$$

and for the infimum over the empty set, one formally defines

$$\inf_{x \in \emptyset} f(x) \;=\; +\infty$$

(where the properties of f then play no role).

Example 1.2.2 We have $\inf_{x \in \mathbb{R}} (x-5) = -\infty$ and $\inf_{x \in \emptyset} (x-5) = +\infty$.

The 'generalized minimal value' $\inf_{x \in M} f(x)$ of P is thus always an element of the *extended real numbers* $\overline{\mathbb{R}} = \mathbb{R} \cup \{\pm\infty\}$. In analysis, it is shown (e.g. [18]), that the infimum exists without any assumptions on f and M and that it is uniquely determined. Furthermore, a characterization of infima is proven there, which we will use subsequently: The infimum of a nonempty set of real numbers is exactly the one of its lower bounds which can be approximated arbitrarily well by elements of the set. For the infima of functions on sets considered here, this means that $v = \inf_{x \in M} f(x)$ holds for $M \ne \emptyset$ if and only if $v \le f(x)$ is true for all $x \in M$ and a sequence $(x^k) \subseteq M$ with $v = \lim_k f(x^k)$ exists. Here and in the following, we write briefly (x^k) for a sequence $(x^k)_{k \in \mathbb{N}}$ and $\lim_k$ for $\lim_{k \to \infty}$.

Definition 1.2.3 (Solvability) The minimization problem P is called *solvable* if some $\bar{x} \in M$ with $\inf_{x \in M} f(x) = f(\bar{x})$ exists.

Solvability of P thus means that the infimum of f on M can be realized as the objective function value of some feasible point, i.e., the infimum is *attained*. To indicate that the infimum is attained we write $\min_{x \in M} f(x)$ instead of $\inf_{x \in M} f(x)$.

Example 1.2.4
We have $0 = \min_{x \in \mathbb{R}} (x - 5)^2 = (\bar{x} - 5)^2$ with $\bar{x} = 5$, but there is no $\bar{x} \in \mathbb{R}$ with $0 = \inf_{x \in \mathbb{R}} e^x = e^{\bar{x}}$.

The following theorem states (unsurprisingly) that solvability is equivalent to the existence of a global minimal point (for a proof see [36]).

Theorem 1.2.5 *The minimization problem P is solvable if and only if it has a global minimal point.*

There are exactly three reasons why P can be unsolvable (in [36] it is shown that there are no further reasons):

- $\inf_{x \in M} f(x) = +\infty$ holds.
  By definition this corresponds to the case $M = \emptyset$, which appears to be trivial, but is not always easy to recognize. For example, if in Example 1.1.1 the additional constraint $r \geq 1$ is introduced (e.g., for tin can marketing reasons), then P_{can} has no feasible points in the case $A < 2\pi$. Sufficient conditions for the solvability of P naturally always require $M \neq \emptyset$.
- $\inf_{x \in M} f(x) = -\infty$ holds.
  In the case of a continuous objective function f, M must be unbounded in this case. For example, the set of feasible points of the optimization problem P_{can} from Example 1.1.1 is unbounded (Exercise 1.2.7). The fact that P_{can} is nevertheless solvable shows that, on the other hand, an unbounded feasible set does not necessarily prevent solvability. As a sufficient condition for solvability, it is still advisable to assume M as bounded. One therefore requires that a radius $R > 0$ can be found such that the ball around the origin with radius R encloses the set M:

$$\exists\, R > 0\, \forall\, x \in M : \quad \|x\| \leq R \text{ (the choice of the norm does not matter).}$$

- A finite infimum $\inf_{x \in M} f(x)$ is not attained.
  The reason for this can again be an unbounded set M, for example with the function $f(x) = e^x$ on the set $M = \mathbb{R}$. But even for bounded sets M this effect is possible, for example when parts of its boundary do not belong to M, as for $f(x) = x$ and $M = (0, 1]$. Here, every feasible point can be replaced by another feasible point with better objective value (Fig. 1.6).

Fig. 1.6 Unsolvability due to lack of closedness of M

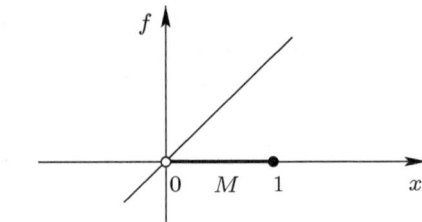

Fig. 1.7 Unsolvability due to jump discontinuity of f

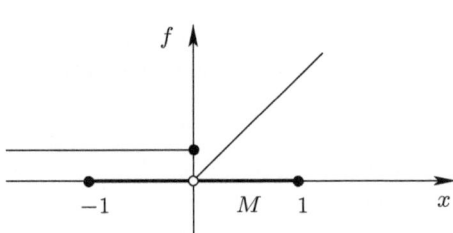

As a remedy, one may assume M to be *closed*, i.e., for all sequences $(x^k) \subseteq M$ with $\lim_k x^k = x^\star$ it holds $x^\star \in M$ (e.g., $M = (0, 1]$ is not closed because $(1/k) \subseteq M$ and $\lim_k(1/k) = 0 \notin M$ hold). If M is described by a finite number of inequality and equality constraints, i.e., for

$$M = \{x \in \mathbb{R}^n | \ g_i(x) \le 0, \ i \in I, \ h_j(x) = 0, \ j \in J\}$$

with finite index sets I and J, then the continuity of the functions $g_i : \mathbb{R}^n \to \mathbb{R}$, $i \in I$, and $h_j : \mathbb{R}^n \to \mathbb{R}$, $j \in J$, is sufficient for the closedness of M (Exercise 1.2.11). If M is simultaneously bounded and closed, then M is called *compact*.

Finally, it is possible that a finite infimum is not attained even on a nonempty and compact set M, namely when f has jump discontinuities. For example, the function

$$f(x) = \begin{cases} 1, & x \le 0 \\ x, & x > 0 \end{cases}$$

does not have a global minimal point on the nonempty and compact set $M = [-1, 1]$, because again every feasible point can be improved by another one (Fig. 1.7). As a remedy, one can assume f to be continuous.

The following fundamental theorem on the existence of minimal and maximal points shows that, under the 'means against unsolvability' motivated above, optimal points do indeed always exist. A version of the theorem, which under a weaker continuity assumption on f still guarantees the existence of minimal points (but not of maximal points), can be found in [34].

Theorem 1.2.6 (Weierstrass Theorem) *Let the set $M \subseteq \mathbb{R}^n$ be nonempty and compact, and let the function $f : M \to \mathbb{R}$ be continuous. Then f possesses (at least) one global minimal point and one global maximal point on M.*

Proof Let $v = \inf_{x \in M} f(x)$. Because of $M \neq \emptyset$ we have $v < +\infty$. It remains to show the existence of some $\bar{x}$ in M with $v = f(\bar{x})$. Since v is the infimum, there exists a sequence $(x^k) \subseteq M$ with $\lim_k f(x^k) = v$. In analysis, it is proven (in the Bolzano-Weierstrass theorem; e.g., [17]), that every sequence (x^k) contained in a compact set M has a subsequence convergent in M. To avoid a tedious subsequence notation, we choose our sequence (x^k) directly as such a convergent sequence, so there exists some $x^\star \in M$ with $\lim_k x^k = x^\star$. The continuity of f on M implies

$$f(x^\star) = f\left(\lim_k x^k\right) = \lim_k f(x^k) = v,$$

so that we can choose $\bar{x} := x^\star$. The proof for the existence of a global maximal point proceeds analogously. $\square$

Although Theorem 1.2.6 has many practical applications, on the other hand its assumptions are violated even for some simple solvable problems, for example in the problem P_{can} from Example 1.1.1.

Exercise 1.2.7 Show that the set of feasible points of the optimization problem P_{can} from Example 1.1.1 is nonempty and closed, but unbounded.

For problems without constraints, so-called *unconstrained problems*, $M = \mathbb{R}^n$ applies (as in Example 1.1.4). While M is then nonempty and closed, it is *not* bounded. Therefore, Theorem 1.2.6 is not applicable to any unconstrained problem.

To make the Weierstrass theorem applicable for problems with an unbounded set M, one uses a trick and considers lower level sets of f. In their definitions as well as some later ones, we will denote the domain of f not with M, but with X, as it does not necessarily have to be the feasible set of some optimization problem.

Definition 1.2.8 (Lower Level Set) For $X \subseteq \mathbb{R}^n$, $f : X \to \mathbb{R}$ and $\alpha \in \mathbb{R}$, we call

$$\text{lev}_{\leq}^{\alpha}(f, X) = \{x \in X | f(x) \leq \alpha\}$$

lower level set of f on X at level α. In the case $X = \mathbb{R}^n$ we also write briefly

$$f_{\leq}^{\alpha} := \text{lev}_{\leq}^{\alpha}(f, \mathbb{R}^n) \quad (= \{x \in \mathbb{R}^n | f(x) \leq \alpha\}).$$

Fig. 1.8 Lower level set $f_\le^1$
of $f(x) = x^2$ on $\mathbb{R}$

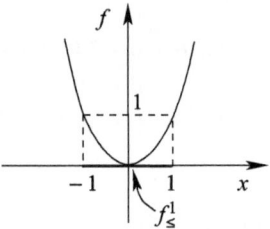

Fig. 1.9 Lower level set $f_\le^1$
of $f(x) = x_1^2 + x_2^2$ on $\mathbb{R}^2$

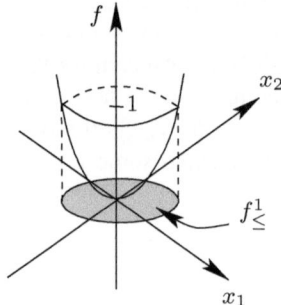

Example 1.2.9 For $f(x) = x^2$ one has $f_\le^1 = [-1, 1]$, $f_\le^0 = \{0\}$ and $f_\le^{-1} = \emptyset$ (Fig. 1.8), and for $f(x) = x_1^2 + x_2^2$ we find $f_\le^1 = \{x \in \mathbb{R}^2 | \; x_1^2 + x_2^2 \le 1\}$ (the unit disk), $f_\le^0 = \{0\}$ and $f_\le^{-1} = \emptyset$ (Fig. 1.9).

Exercise 1.2.10 For a closed set $X \subseteq \mathbb{R}^n$, let the function $f : X \to \mathbb{R}$ be continuous. Show that then for each $\alpha \in \mathbb{R}$ the set $\operatorname{lev}_\le^\alpha(f, X)$ is closed.

Exercise 1.2.11 For a closed set $X \subseteq \mathbb{R}^n$ and finite index sets I and J, let the functions $g_i : X \to \mathbb{R}, i \in I$, and $h_j : X \to \mathbb{R}, j \in J$, be continuous. Show that then the set

$$M \; = \; \{x \in X | \; g_i(x) \le 0, \; i \in I, \; h_j(x) = 0, \; j \in J\}$$

is closed.

For the following result, we introduce the set of global minimal points

$$S \; = \; \{\bar{x} \in M | \; \forall x \in M : \; f(x) \ge f(\bar{x})\}$$

of P. The solvability of P can then be expressed by the condition $S \ne \emptyset$ (in [36] even more interesting properties of the set S are shown).

Lemma 1.2.12 *For some $\alpha \in \mathbb{R}$ let $\text{lev}_{\leq}^{\alpha}(f, M) \neq \emptyset$. Then $S \subseteq \text{lev}_{\leq}^{\alpha}(f, M)$ holds.*

Proof In view of $\text{lev}_{\leq}^{\alpha}(f, M) \neq \emptyset$ there exists some point $\tilde{x}$ in M with $f(\tilde{x}) \leq \alpha$. Then every global minimal point $\bar{x}$ of P satisfies $\bar{x} \in M$ and $f(\bar{x}) \leq f(\tilde{x}) \leq \alpha$, so $\bar{x} \in \text{lev}_{\leq}^{\alpha}(f, M)$. $\qquad\square$

The concept of lower level sets allows us to consider the interplay of the properties of the objective function f and the feasible set M in the following sufficient condition for the solvability of P.

Theorem 1.2.13 (Strengthened Weierstrass Theorem) *For a (not necessarily bounded or closed) set $M \subseteq \mathbb{R}^n$, let $f : M \to \mathbb{R}$ be continuous, and with some $\alpha \in \mathbb{R}$, let $\text{lev}_{\leq}^{\alpha}(f, M)$ be nonempty and compact. Then f possesses (at least) one global minimal point on M.*

Proof Due to Lemma 1.2.12, P and the auxiliary problem

$$\tilde{P} : \quad \min f(x) \quad \text{s.t.} \quad x \in \text{lev}_{\leq}^{\alpha}(f, M)$$

have the same minimal points and the same minimal value. Since $\tilde{P}$ fulfills the assumptions of Theorem 1.2.6, the assertion follows. $\qquad\square$

Exercise 1.2.14 Show that the assumptions of Theorem 1.2.13 are weaker than those of Theorem 1.2.6, that is, they can be fulfilled under the assumptions of Theorem 1.2.6.

The *strengthening* of Theorem 1.2.13 compared to Theorem 1.2.6 refers to the statement of interest to us from the Weierstrass theorem, namely the existence of a global *minimal* point, which also follows under the weaker assumptions of Theorem 1.2.13. Since there, however, no statement can be made about the existence of a global *maximal* point of P, the two theorems are in fact independent of each other.

Example 1.2.15 Consider the problem

$$P : \quad \min e^x \quad \text{s.t.} \quad x \geq 0$$

(Fig. 1.10).

Fig. 1.10 e^x with $x \geq 0$

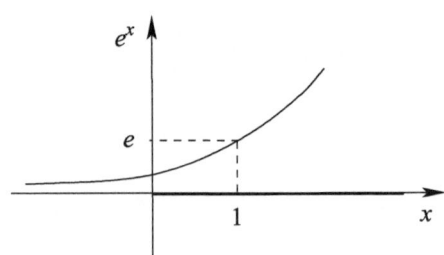

Here, $M = \{x \in \mathbb{R} | x \geq 0\}$ is unbounded, so that Theorem 1.2.6 is not applicable. However, for example with $\alpha = e$ the set

$$\mathrm{lev}_{\leq}^e(f, M) = \{x \in M | e^x \leq e\} = \{x \geq 0 | x \leq 1\} = [0, 1]$$

is nonempty and compact. Consequently, Theorem 1.2.13 is applicable and P therefore solvable.

Exercise 1.2.16 Show the solvability of the optimization problem P_{can} from Example 1.1.1 using Theorem 1.2.13.

> **Corollary 1.2.17 (Strengthened Weierstrass Theorem for Unconstrained Problems)** *Let the function $f : \mathbb{R}^n \to \mathbb{R}$ be continuous and, with some $\alpha \in \mathbb{R}$, let $f_{\leq}^\alpha$ be nonempty and compact. Then f has (at least) one global minimal point on $\mathbb{R}^n$.*

Proof Theorem 1.2.13 with $M = \mathbb{R}^n$. □

Example 1.2.18 For $f(x) = (x - 5)^2$ the set $f_{\leq}^1 = [4, 6]$ is nonempty and compact, so f has a global minimal point on $\mathbb{R}$ according to Corollary 1.2.17.

Example 1.2.19 With $f(x) = e^x$ we obtain $f_{\leq}^\alpha = \emptyset$ for all $\alpha \leq 0$ and $f_{\leq}^\alpha = (-\infty, \log(\alpha)]$ for all $\alpha > 0$. Therefore, for no α the set $f_{\leq}^\alpha$ is nonempty and compact, and Corollary 1.2.17 is not applicable. Indeed, f does not have a global minimal point on $\mathbb{R}$.

Example 1.2.20 Also for $f(x) = \sin(x)$ Corollary 1.2.17 is not applicable, as all sets $f_{\leq}^\alpha$ with $\alpha \in \mathbb{R}$ are unbounded or empty. However, f does possess global minimal points on $\mathbb{R}$.

In the following, we provide a simple criterion from which the compactness of $\text{lev}^\alpha_\leq(f, X)$ follows for *every* $\alpha \in \mathbb{R}$. This allows us to guarantee the assumptions of Theorem 1.2.13 and Corollary 1.2.17 without having to specify an explicit level α.

Definition 1.2.21 (Coercivity) Let a closed set $X \subseteq \mathbb{R}^n$ and a function $f : X \to \mathbb{R}$ be given. If for all sequences $(x^k) \subseteq X$ with $\lim_k \|x^k\| = +\infty$ also

$$\lim_k f(x^k) = +\infty$$

holds, then f is called *coercive* on X.

Example 1.2.22 The function $f(x) = (x - 5)^2$ is coercive on $\mathbb{R}$.

Example 1.2.23 The function $f(x) = e^x$ is not coercive on $X = \mathbb{R}$, but it is on the set $X = \{x \in \mathbb{R} | x \geq 0\}$.

Exercise 1.2.24 Let the quadratic function $q(x) = \frac{1}{2}x^\mathsf{T} A x + b^\mathsf{T} x$ be given with a symmetric (n, n)-matrix A (i.e., it holds $A = A^\mathsf{T}$) and $b \in \mathbb{R}^n$. Show that q is coercive on $\mathbb{R}^n$ if and only if A is positive definite (i.e., if $d^\mathsf{T} A d > 0$ holds for all $d \in \mathbb{R}^n \setminus \{0\}$; details on positive definite matrices can be found in Sect. 2.1.4).

Example 1.2.25 On compact sets X every function f is trivially coercive.

Regarding the formulation of Example 1.2.25 we emphasize that the term 'trivial' is used carefully in this textbook. It does not refer to statements that are 'easy' to prove from the author's point of view, but to those that are valid due to a logical triviality. For example, the statement 'all green cows can fly' is trivially true, because to refute it, one would have to find a green cow that cannot fly. But since one cannot find a green cow in the first place, there is no need to look for a green cow that cannot fly. Therefore, the statement cannot be refuted for a trivial reason and is consequently true. In Example 1.2.25, the analogous argument is that in a compact set X, there is not a single sequence (x^k) with $\lim_k \|x^k\| \to +\infty$. To show that f is *not* coercive, however, such a sequence would have to *exist* and also fail to satisfy $\lim_k f(x^k) = +\infty$. The latter is irrelevant, however, because already the existence of the sequence is ruled out. Consequently, f is coercive on X for a trivial reason.

With these concepts, in the Weierstrass theorem we can replace the boundedness of M by the coercivity of f on M in the sense of the following two results. Proofs can be found, for example, in [36].

Lemma 1.2.26 *Let the function $f : X \to \mathbb{R}$ be continuous and coercive on the (not necessarily bounded) closed set $X \subseteq \mathbb{R}^n$. Then the set $\text{lev}^\alpha_\leq(f, X)$ is compact for every level $\alpha \in \mathbb{R}$.*

From this and from the Weierstrass theorem follows the second result, which in particular provides a frequently applicable criterion for proving the solvability of unconstrained optimization problems.

Corollary 1.2.27 *Let M be nonempty and closed, but not necessarily bounded. Furthermore, let the function $f : M \to \mathbb{R}$ be continuous and coercive on M. Then f possesses (at least) one global minimal point on M.*

1.3   Rules of Calculus and Transformations

This section introduces a series of rules of calculus and transformations of optimization problems that are of interest in this textbook. The existence of all occurring optimal points and values is assumed in this section without further mention and must be guaranteed, for example, with the techniques from Sect. 1.2. The transfer of the results on optimal values to cases of nonattained infima and suprema is left to the reader as further exercise.

Exercise 1.3.1 (Scalar Multiples and Sums) Let $M \subseteq \mathbb{R}^n$ and $f, g : M \to \mathbb{R}$. Then the following holds:

(a) $\forall \alpha \geq 0, \ \beta \in \mathbb{R} : \ \min_{x \in M} (\alpha f(x) + \beta) = \alpha (\min_{x \in M} f(x)) + \beta$.
(b) $\forall \alpha < 0, \ \beta \in \mathbb{R} : \ \min_{x \in M} (\alpha f(x) + \beta) = \alpha (\max_{x \in M} f(x)) + \beta$.
(c) $\min_{x \in M} (f(x) + g(x)) \geq \min_{x \in M} f(x) + \min_{x \in M} g(x)$.
(d) In statement c, the strict inequality > can occur.

In statement a and statement b, the local and global optimal points, respectively, of the optimization problems also coincide.

Exercise 1.3.2 (Separable Objective Function on Cartesian Product) Let $X \subseteq \mathbb{R}^n, Y \subseteq \mathbb{R}^m, f : X \to \mathbb{R}$ and $g : Y \to \mathbb{R}$. Then it holds

$$\min_{(x,y) \in X \times Y} (f(x) + g(y)) = \min_{x \in X} f(x) + \min_{y \in Y} g(y).$$

Exercise 1.3.3 (Swapping Minimizations and Maximizations) Let $X \subseteq \mathbb{R}^n, Y \subseteq \mathbb{R}^m, M = X \times Y$ and $f : M \to \mathbb{R}$. Then it holds:

(a) $\min_{(x,y) \in M} f(x, y) = \min_{x \in X} \min_{y \in Y} f(x, y) = \min_{y \in Y} \min_{x \in X} f(x, y)$.
(b) $\max_{(x,y) \in M} f(x, y) = \max_{x \in X} \max_{y \in Y} f(x, y) = \max_{y \in Y} \max_{x \in X} f(x, y)$.
(c) $\min_{x \in X} \max_{y \in Y} f(x, y) \geq \max_{y \in Y} \min_{x \in X} f(x, y)$.
(d) In statement c, the strict inequality > can occur.

Exercise 1.3.4 (Monotone Transformation) For $M \subseteq \mathbb{R}^n$ and a function $f : M \to Y$ with $Y \subseteq \mathbb{R}$, let $\psi : Y \to \mathbb{R}$ be a strictly monotonically increasing function. Then it holds

$$\min_{x \in M} \psi(f(x)) = \psi(\min_{x \in M} f(x)),$$

and the local and global minimal points, respectively, coincide.

Exercise 1.3.5 (Epigraph Reformulation) Consider $M \subseteq \mathbb{R}^n$ and a function $f : M \to \mathbb{R}$. Then the problems

$$P : \qquad \min_{x \in \mathbb{R}^n} f(x) \quad \text{s.t.} \quad x \in M$$

and

$$P_{\text{epi}} : \qquad \min_{(x,\alpha) \in \mathbb{R}^n \times \mathbb{R}} \alpha \quad \text{s.t.} \quad f(x) \leq \alpha, \quad x \in M$$

are equivalent in the following sense:

(a) For every local or global minimal point $x^\star$ of P, $(x^\star, f(x^\star))$ is a local or global minimal point of P_{epi}, respectively.
(b) For every local or global minimal point $(x^\star, \alpha^\star)$ of P_{epi}, $x^\star$ is a local or global minimal point of P, respectively.
(c) The minimal values of P and P_{epi} coincide.

Definition 1.3.6 (Parallel Projection) Let $M \subseteq \mathbb{R}^n \times \mathbb{R}^m$. Then

$$\mathrm{pr}_x M = \{x \in \mathbb{R}^n \mid \exists y \in \mathbb{R}^m : (x, y) \in M\}$$

is called the *parallel projection* of M onto (the 'x-space') $\mathbb{R}^n$.

Exercise 1.3.7 (Projection Reformulation) Consider $M \subseteq \mathbb{R}^n \times \mathbb{R}^m$ and a function $f : \mathbb{R}^n \to \mathbb{R}$, which does not depend on the variables from $\mathbb{R}^m$. Then the problems

$$P : \qquad \min_{(x,y) \in \mathbb{R}^n \times \mathbb{R}^m} f(x) \quad \text{s.t.} \quad (x, y) \in M$$

and

$$P_{\text{proj}} : \qquad \min_{x \in \mathbb{R}^n} f(x) \quad \text{s.t.} \quad x \in \mathrm{pr}_x M$$

are equivalent in the following sense:

(a) For every local or global minimal point $(x^\star, y^\star)$ of P, $x^\star$ is a local or global minimal point of P_{proj}, respectively.
(b) For every local or global minimal point $x^\star$ of P_{proj}, there exists some $y^\star \in \mathbb{R}^m$ such that $(x^\star, y^\star)$ is a local or global minimal point of P, respectively.
(c) The minimal values of P and P_{proj} coincide.

Unconstrained Optimization

2

Contents

Nonlinear optimization problems without constraints have the form

$$P: \quad \min \ f(x)$$

with a function $f : \mathbb{R}^n \to \mathbb{R}$. The problem P is then called *unconstrained optimization problem*. More generally, sometimes problems with an *open* feasible set M are referred to as unconstrained (roughly speaking, because no 'boundary effects' can occur in these problems). Indeed, the derivative-based optimality conditions discussed in the following Sect. 2.1 can easily be transferred to this case from which, however, we will refrain for the sake of clarity. Section 2.2 discusses various algorithmic concepts for the solution of unconstrained optimization problems, which are based on the previously derived optimality conditions.

© The Editor(s) (if applicable) and The Author(s), under exclusive license   21
to Springer-Verlag GmbH, DE, part of Springer Nature 2024
O. Stein, *Basic Concepts of Nonlinear Optimization*,
Mathematics Study Resources 8, https://doi.org/10.1007/978-3-662-69741-2_2

2.1 Optimality Conditions

To develop derivative-based necessary and sufficient optimality conditions, we first introduce in Sect. 2.1.1 the derivative-free concept of a descent direction for a function f at a point $\bar{x} \in \mathbb{R}^n$. For every differentiable function f, using the (multidimensional) first derivative of f at $\bar{x}$, a sufficient condition for a direction to be descent direction for f at $\bar{x}$ can be given. From the absence of such descent directions we derive in Sect. 2.1.2 the central necessary optimality condition, Fermat's rule. Since only first derivatives are used, one speaks of a first-order sufficient condition for descent, and of a first-order necessary optimality condition.

The thus employed definition of the multidimensional first derivative leads to the concept of the gradient of the function f at $\bar{x}$, which itself is a vector of length n. The geometric properties of gradients are discussed in Sect. 2.1.3. They are fundamental for understanding both the optimality conditions and the algorithms in Sect. 2.2.

Provided that f is even twice differentiable, the above first-order sufficient condition for the descent property of a direction can be refined by information about the second derivative of f at $\bar{x}$, which provides a stronger necessary optimality condition than Fermat's rule, namely the second-order necessary optimality condition discussed in Sect. 2.1.4. By a simple modification, this concept can also be used to construct a sufficient optimality condition. However, there exists a 'gap' between the second-order necessary and sufficient optimality conditions. We briefly discuss why this does not have serious consequences in practice. The concluding Sect. 2.1.5 outlines how the optimality conditions simplify if the objective function f is additionally convex on $\mathbb{R}^n$.

2.1.1 Descent Directions

To determine conditions that a function f must necessarily satisfy at a minimal point, we proceed according to the following exclusion principle: If one can 'move away' from the point $\bar{x} \in \mathbb{R}^n$ along a direction $d \in \mathbb{R}^n$ while the function values decrease (at least initially), then $\bar{x}$ cannot be a local minimal point. The points that one visits when leaving $\bar{x}$ along d can be explicitly addressed as $\bar{x} + td$ with scalars $t \geq 0$, i.e., using the parametric form of a straight line.

> **Definition 2.1.1 (Descent Direction)** Let $f : \mathbb{R}^n \to \mathbb{R}$ and $\bar{x} \in \mathbb{R}^n$. A vector $d \in \mathbb{R}^n$ is called *descent direction* for f in $\bar{x}$, if
>
> $$\exists \check{t} > 0 \quad \forall t \in (0, \check{t}) : \quad f(\bar{x} + td) < f(\bar{x})$$
>
> holds.

Exercise 2.1.2 For $f : \mathbb{R}^n \to \mathbb{R}$, let $\bar{x}$ be a local minimal point. Show that then no descent direction for f in $\bar{x}$ exists.

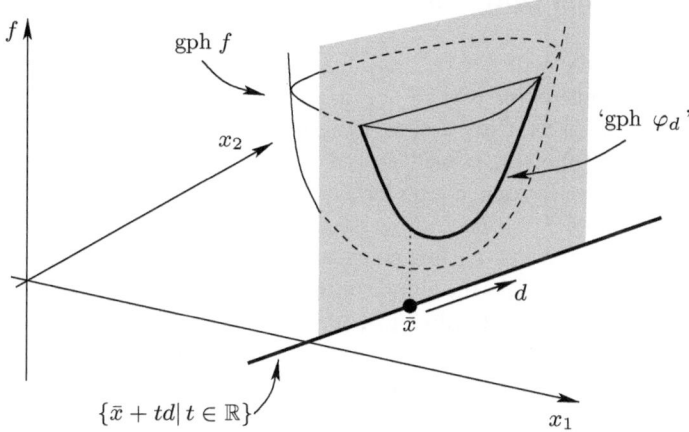

Fig. 2.1 One-dimensional restriction

In the following, we will examine the function $f(\bar{x} + td)$ of the one-dimensional variable t and, to this end, introduce a name for this univariate object.

Definition 2.1.3 (One-Dimensional Restriction) Consider $f : \mathbb{R}^n \rightarrow \mathbb{R}$, a point $\bar{x} \in \mathbb{R}^n$ and a direction vector $d \in \mathbb{R}^n$. The univariate function

$$\varphi_d : \mathbb{R}^1 \rightarrow \mathbb{R}^1, \quad t \mapsto f(\bar{x} + td)$$

is called *one-dimensional restriction* of f to the line passing through $\bar{x}$ in direction d.

Figure 2.1 illustrates how the graph of the function φ_d can be geometrically thought of to be derived from that of the function f for $n = 2$: The point $\bar{x}$ and the direction d define the line $\{\bar{x} + td \mid t \in \mathbb{R}\}$ in the two-dimensional space of arguments, and φ_d describes the evaluation of the function f at the points of this line. Geometrically, the graph gph φ_d of φ_d is obtained by 'erecting a vertical plane' over the line and intersecting it with the graph gph$f$ of f. Of course, gph φ_d is not a subset of $\mathbb{R}^3$, as depicted in Fig. 2.1, but a subset of the two-dimensional space that coincides with the constructed 'vertical plane'.

Since $\varphi_d(0) = f(\bar{x})$ holds for every direction $d \in \mathbb{R}^n$, the latter is a descent direction for f in $\bar{x}$ if and only if

$$\exists \check{t} > 0 \quad \forall t \in (0, \check{t}) : \quad \varphi_d(t) < \varphi_d(0)$$

holds.

2.1.2 First-Order Optimality Condition

In the following, we will derive a condition that must necessarily be fulfilled at every local minimal point of f. Since it uses information from first derivatives of f, it is referred to as a *first-order necessary optimality condition*. To allow an application of the following considerations also to certain *non*smooth problems (in Theorem 3.3.21), we introduce it under a rather weak assumption on the function f, namely its *one-sided directional differentiability*.

To motivate this concept, we first consider the case in which f is differentiable at a considered point $\bar{x} \in \mathbb{R}^n$. As the composition of f with the affine-linear function $\bar{x} + td$, the one-dimensional restriction φ_d is then differentiable at $\bar{t} = 0$. Its derivative $\varphi'_d(0)$ indicates with which slope the function values of f change, when moving away from $\bar{x}$ in direction d. This derivative

$$\varphi'_d(0) \;=\; \lim_{t \to 0} \frac{f(\bar{x} + td) - f(\bar{x})}{t}$$

is called *directional derivative* of f at $\bar{x}$ in direction d.

It seems plausible that in the case $\varphi'_d(0) < 0$ the function values of φ_d initially drop as t increases and, therefore, d is a descent direction. Since *negative* values of t do not play any role for this argument, one actually does not need the directional derivative of f, but only the following concept.

Definition 2.1.4 (One-Sided Directional Derivative) A function $f : \mathbb{R}^n \to \mathbb{R}$ is called *one-sided directionally differentiable* at $\bar{x} \in \mathbb{R}^n$ in a direction $d \in \mathbb{R}^n$, if the limit

$$f'(\bar{x}, d) \;:=\; \lim_{t \searrow 0} \frac{f(\bar{x} + td) - f(\bar{x})}{t}$$

exists. The value $f'(\bar{x}, d)$ is then called *one-sided directional derivative*. The function f is called *one-sided directionally differentiable* at $\bar{x}$, if f is one-sided directionally differentiable at $\bar{x}$ in every direction $d \in \mathbb{R}^n$, and f is called *one-sided directionally differentiable*, if f is one-sided directionally differentiable at every $\bar{x} \in \mathbb{R}^n$.

The above notations for the occurring limits more explicitly mean the following: Directional differentiability of f at $\bar{x}$ in direction d states that for all sequences $(t^k) \subseteq \mathbb{R} \setminus \{0\}$ with $\lim_k t^k = 0$ the limit

$$\lim_k \frac{f(\bar{x} + t^k d) - f(\bar{x})}{t^k}$$

exists and is identical. We call this limit $\varphi_d'(0)$. On the other hand, the *one-sided* directional differentiability of f at $\bar{x}$ in direction d means that only for all sequences $(t^k) \subseteq \{t \in \mathbb{R} \mid t > 0\}$ with $\lim_k t^k = 0$ the limit

$$\lim_k \frac{f(\bar{x} + t^k d) - f(\bar{x})}{t^k}$$

needs to exist and to be identical. We denote this limit by $f'(\bar{x}, d)$.

Every function f that is directionally differentiable at $\bar{x}$ in direction d is also one-sided directionally differentiable with $f'(\bar{x}, d) = \varphi_d'(0)$. Therefore, in particular, every function f that is differentiable at $\bar{x}$ is also one-sided directionally differentiable there. However, the class of one-sided directionally differentiable functions also includes many nondifferentiable functions, such as all convex functions on $\mathbb{R}^n$ (Sect. 2.1.5 and [33]) as well as maxima of a finite number of smooth functions (Exercise 2.1.12). As a simple example, $f(x) = |x|$ is neither differentiable nor directionally differentiable, but *one-sided* directionally differentiable at $\bar{x} = 0$.

Lemma 2.1.5 *Let the function* $f : \mathbb{R}^n \to \mathbb{R}$ *be one-sided directionally differentiable at* $\bar{x} \in \mathbb{R}^n$ *in direction* $d \in \mathbb{R}^n$ *with* $f'(\bar{x}, d) < 0$. *Then d is a descent direction for f at $\bar{x}$.*

Proof Assume that d is not a descent direction for f at $\bar{x}$. Then there exists no $\check{t} > 0$ such that for all $t \in (0, \check{t})$ the inequality $f(\bar{x} + td) < f(\bar{x})$ is fulfilled. In particular, for each $k \in \mathbb{N}$ at least one $t^k \in (0, 1/k)$ satisfies the reverse inequality $f(\bar{x} + t^k d) \geq f(\bar{x})$. Due to $t^k > 0$, $k \in \mathbb{N}$, this implies

$$\frac{f(\bar{x} + t^k d) - f(\bar{x})}{t^k} \geq 0,$$

and the one-sided directional differentiability of f at $\bar{x}$ in direction d yields

$$f'(\bar{x}, d) = \lim_k \frac{f(\bar{x} + t^k d) - f(\bar{x})}{t^k} \geq 0.$$

However, this contradicts the assumption $f'(\bar{x}, d) < 0$. Therefore, the assumption is false and the claim is proven.   □

From Exercise 2.1.2 and Lemma 2.1.5 the next result follows.

Lemma 2.1.6 *Let the function* $f : \mathbb{R}^n \to \mathbb{R}$ *be one-sided directionally differentiable at a local minimal point* $\bar{x} \in \mathbb{R}^n$. *Then* $f'(\bar{x}, d) \geq 0$ *holds for every direction $d \in \mathbb{R}^n$.*

Lemmas 2.1.5 and 2.1.6 motivate the following definitions.

Definition 2.1.7 (First-Order Descent Direction) For a function $f : \mathbb{R}^n \to \mathbb{R}$ that is one-sided directionally differentiable at $\bar{x} \in \mathbb{R}^n$ in direction $d \in \mathbb{R}^n$, d is called a *first-order descent direction*, if $f'(\bar{x}, d) < 0$ holds.

Definition 2.1.8 (Stationary Point—Unconstrained Case) Let the function $f : \mathbb{R}^n \to \mathbb{R}$ be one-sided directionally differentiable at $\bar{x} \in \mathbb{R}^n$. Then $\bar{x}$ is called a *stationary point* of f, if $f'(\bar{x}, d) \geq 0$ holds for every direction $d \in \mathbb{R}^n$.

In this terminology, Lemma 2.1.5 states that every first-order descent direction is indeed a descent direction in the sense of Definition 2.1.1. The definition of stationarity of a point from Definition 2.1.8 states that no first-order descent direction exists at it, and Lemma 2.1.6 states that every local minimal point of a function, which is one-sided directionally differentiable there, is also stationary.

To be able to use Lemma 2.1.6 algorithmically, we need a simple formula for $f'(\bar{x}, d)$ and therefore, for the major part of this textbook, return to a function f that is differentiable at $\bar{x}$. As a direction vector, we first consider the i-th unit vector $d = e_i$ (thus setting $d_i = 1$ and $d_j = 0$ for all $j \neq i$). Then $f'(\bar{x}, e_i)$ is the *partial derivative* of f with respect to the variable x_i, which is alternatively denoted by $\partial_{x_i} f(\bar{x})$. The *first derivative* of a partially differentiable function $f : \mathbb{R}^n \to \mathbb{R}$ at $\bar{x}$ is the row vector

$$Df(\bar{x}) := (\, \partial_{x_1} f(\bar{x}), \, \ldots \, , \partial_{x_n} f(\bar{x}) \,).$$

Its transpose, the column vector $\nabla f(\bar{x}) := (Df(\bar{x}))^\mathsf{T}$, is called the *gradient* of f at $\bar{x}$. With regard to later constructions, we note that the gradient $\nabla f(\bar{x})$, while on the one hand only a list of partial derivative information, can on the other hand be interpreted as an n-dimensional vector, just like, e.g., $\bar{x}$.

For a *vector*-valued function $f : \mathbb{R}^n \to \mathbb{R}^m$ with partially differentiable components $f_1, \ldots, f_m$, the first derivative is defined as

$$Df(\bar{x}) := \begin{pmatrix} Df_1(\bar{x}) \\ \vdots \\ Df_m(\bar{x}) \end{pmatrix}.$$

This (m, n)-matrix is called the *Jacobian matrix* or *derivative matrix* of f at $\bar{x}$. Occasionally, we also use the notation $\nabla f(\bar{x}) := (Df(\bar{x}))^\mathsf{T}$ for vector-valued functions f. In this case, $\nabla f(\bar{x})$ is not called gradient, but transposed Jacobian.

An important rule of calculation for differentiable functions is the *chain rule*, the proof of which can be found, for example, in [17].

Theorem 2.1.9 (Chain Rule) *Let $g : \mathbb{R}^n \to \mathbb{R}^m$ be differentiable at $\bar{x} \in \mathbb{R}^n$ and $f : \mathbb{R}^m \to \mathbb{R}^k$ be differentiable at $g(\bar{x}) \in \mathbb{R}^m$. Then $f \circ g : \mathbb{R}^n \to \mathbb{R}^k$ is differentiable at $\bar{x}$ with*

$$D(f \circ g)(\bar{x}) = Df(g(\bar{x})) \cdot Dg(\bar{x}).$$

A major reason for defining the Jacobian matrix of a function as above (and not, e.g., as its transpose, or a vector of length mn) is that the chain rule can then be formulated completely analogously to the one-dimensional case ($n = m = k = 1$), even though the resulting product is a matrix product.

When applying the chain rule to the function $\varphi_d(t) = f(\bar{x} + td)$, we have $k = m = 1$ and $g(t) = \bar{x} + td$. The Jacobian matrix of g is

$$Dg(t) = d,$$

and thus

$$\varphi_d'(0) = Df(\bar{x}) d$$

follows. In this specific case, the matrix product from the chain rule becomes the product of the row vector $Df(\bar{x})$ with the column vector d.

For two general (column) vectors $a, b \in \mathbb{R}^n$, the term defined as

$$a^\mathsf{T} b = \sum_{i=1}^n a_i b_i$$

is called the (standard) *inner product* of a and b. An alternative notation for this is

$$\langle a, b \rangle := a^\mathsf{T} b.$$

We thus obtain

$$\varphi_d'(0) = \langle \nabla f(\bar{x}), d \rangle$$

and may formulate Lemma 2.1.5 for the case of a differentiable function f.

Lemma 2.1.10 *Let the function* $f : \mathbb{R}^n \to \mathbb{R}$ *be differentiable at the point* $\bar{x} \in \mathbb{R}^n$, *and for the direction* $d \in \mathbb{R}^n$ *let* $\langle \nabla f(\bar{x}), d \rangle < 0$ *hold. Then* d *is a descent direction for* f *at* $\bar{x}$.

For a function f that is differentiable at $\bar{x}$, d is a first-order descent direction in the sense of Definition 2.1.7 if $\langle \nabla f(\bar{x}), d \rangle < 0$ holds.

Remark 2.1.11 It will be important to interpret the condition $\langle \nabla f(\bar{x}), d \rangle < 0$ geometrically. For two vectors $a, b \in \mathbb{R}^n$, the inner product $\langle a, b \rangle$ has, in addition to the algebraic definition as $a^\mathsf{T} b$, also the representation

$$\langle a, b \rangle \;=\; \|a\|_2 \cdot \|b\|_2 \cdot \cos(\angle(a, b)) \tag{2.1}$$

(e.g., [8]). Here, $\angle(a, b)$ denotes the angle between the two vectors a and b. This can be defined for n-dimensional vectors a and b by measuring it in the plane spanned by a and b (i.e., in the set $\{\lambda a + \mu b \,|\, \lambda, \mu \in \mathbb{R}\}$). In the exceptional case where the vectors a and b are linearly dependent, they do not span a plane, but the angle between a and b can only be zero (if they point in the same direction) or π (if they point in opposite directions). In these cases, $\cos(\angle(a, b)) = +1$ or $\cos(\angle(a, b)) = -1$ hold, respectively.

From (2.1) we obtain that $\langle a, b \rangle < 0$ holds if and only if $\cos(\angle(a, b)) < 0$ is true, i.e., exactly for $\angle(a, b) \in (\pi/2, 3\pi/2)$. In other words, the inner product of vectors a and b is negative if and only if they form an obtuse angle. Similarly, the inner product is positive for vectors forming an acute angle, and zero for vectors perpendicular to each other.

In particular, d is a first-order descent direction for f at $\bar{x}$ if and only if d forms an *obtuse angle* with the gradient $\nabla f(\bar{x})$. We will later see that, under certain additional conditions, directions d *perpendicular* to $\nabla f(\bar{x})$ may also be descent directions.

Even for functions that are only one-sided directionally differentiable, simple formulas for the one-sided directional derivative can sometimes be given.

Exercise 2.1.12 Consider $\bar{x} \in \mathbb{R}^n$, a finite index set K and functions $f_k : \mathbb{R}^n \to \mathbb{R}$ that are differentiable at $\bar{x}$, $k \in K$. Show that then the function $f(x) := \max_{k \in K} f_k(x)$ is one-sided directionally differentiable at $\bar{x}$ and that with $K_\star(\bar{x}) = \{k \in K \,|\, f_k(\bar{x}) = f(\bar{x})\}$

$$f'(\bar{x}, d) \;=\; \max_{k \in K_\star(\bar{x})} \langle \nabla f_k(\bar{x}), d \rangle$$

holds for every direction $d \in \mathbb{R}^n$.

We can now prove the central optimality condition for unconstrained smooth optimization problems.

> **Theorem 2.1.13 (First-Order Necessary Optimality Condition—Fermat's Rule)** *Let the function* $f : \mathbb{R}^n \to \mathbb{R}$ *be differentiable at a local minimal point* $\bar{x} \in \mathbb{R}^n$. *Then* $\nabla f(\bar{x}) = 0$ *holds.*

Proof According to Lemma 2.1.6, $\bar{x}$ is a stationary point of f. Due to the representation of the directional derivative by the chain rule, for every direction $d \in \mathbb{R}^n$ we have

$$0 \leq \varphi_d'(0) = \langle \nabla f(\bar{x}), d \rangle.$$

In particular, this applies to the choice $d := -\nabla f(\bar{x})$, which yields

$$0 \leq \langle \nabla f(\bar{x}), -\nabla f(\bar{x}) \rangle = -\|\nabla f(\bar{x})\|_2^2 \leq 0.$$

It follows $\|\nabla f(\bar{x})\|_2 = 0$ and, due to the definiteness of the norm, $\nabla f(\bar{x}) = 0$. $\quad\square$

Fermat's rule is referred to as a *first-order optimality condition* since it makes use of first derivatives of the function f. It motivates the following definition.

> **Definition 2.1.14 (Critical Point)** Let the function $f : \mathbb{R}^n \to \mathbb{R}$ be differentiable at $\bar{x} \in \mathbb{R}^n$. Then $\bar{x}$ is called a *critical point* of f if $\nabla f(\bar{x}) = 0$ holds.

In this terminology, according to Fermat's rule, every local minimal point of a differentiable function is necessarily a critical point.

Exercise 2.1.15 Let the function $f : \mathbb{R}^n \to \mathbb{R}$ be differentiable at a point $\bar{x} \in \mathbb{R}^n$. Show that $\bar{x}$ is a stationary point of f if and only if it is a critical point of f.

Exercise 2.1.15 justifies why in the literature on smooth unconstrained optimization the terms *stationary point* and *critical point* are used synonymously. For, e.g., nonsmooth or constrained smooth problems, however, the relationship between stationarity defined by the absence of a first-order descent direction and an algebraic optimality condition (analogous to criticality $\nabla f(\bar{x}) = 0$) is less clear ([33] and Chap. 3). Nevertheless, the terms *stationarity* and *criticality* are not used consistently in the literature.

Example 2.1.16 For the (global) minimal point $\bar{x} = 0$ of $f_1(x) = x_1^2 + x_2^2$, one calculates $\nabla f_1(\bar{x}) = 0$. For the two functions $f_2(x) = -x_1^2 - x_2^2$ and $f_3(x) = x_1^2 - x_2^2$, $\bar{x} = 0$ is also a critical point, although $\bar{x}$ is *not* a local minimal point.

Indeed, the direction $d = (0, 1)^\mathsf{T}$ yields for the one-dimensional restriction of both functions

$$\varphi_d(t) \;=\; f_i(\bar{x} + td) \;=\; f_i(0, t) \;=\; -t^2, \qquad i = 2, 3,$$

so that, while leaving $\bar{x} = 0$ in this direction, the function values decrease (thus d is a descent direction, but *not* a first-order descent direction). For f_2, this is even the case in every direction, as f_2 has a maximal point at $\bar{x} = 0$. For f_3, however, the function values increase in the direction $d = (1, 0)^\mathsf{T}$. In this case, one speaks of a saddle point of f_3.

> **Definition 2.1.17 (Saddle Point)** Let the function $f : \mathbb{R}^n \to \mathbb{R}$ be differentiable at $\bar{x} \in \mathbb{R}^n$. Then $\bar{x}$ is called a *saddle point* of f, if it is a critical point, but neither a local minimal nor a local maximal point of f.

Example 2.1.16 illustrates that Fermat's rule is only a *necessary* condition for optimality. It states that a local minimal point of f necessarily is a critical point, but the property of being a critical point is not *sufficient* for minimality.

This makes it clear that critical points are merely *candidates* for minimal points of f, but can also correspond to maximal or saddle points. Algorithm 2.1 describes a conceptual procedure for minimization using critical points, based on this observation. It uses the necessary condition for optimality to 'filter out' all feasible points (i.e., all points in $\mathbb{R}^n$) that do not qualify as candidates for minimal points. 'Only' among the remaining points a minimal point must then be sought.

Here and in the following, we call an optimization problem P differentiable if it is described by differentiable functions. In the present unconstrained case, this just concerns the differentiability of the objective function f.

Algorithm 2.1: Conceptual algorithm for unconstrained nonlinear minimization with first-order information

---

    Input: Solvable unconstrained differentiable optimization problem P
    Output: Global minimal point $x^\star$ of f over $\mathbb{R}^n$

1 begin
2   Determine all critical points of f, i.e. the solution set C of the equation $\nabla f(x) = 0$.
3   Determine a minimal point $x^\star$ of f in C.
4 end

Algorithm 2.1 has three disadvantages that impede its application to practical problems. First, the solvability of the problem P must be known a priori, for example by applying the criteria from Sect. 1.2. For example, the function $f(x) = x^3/3 - x$ has exactly the two critical points $x^1 = -1$ and $x^2 = 1$, where x^2 has the smaller function value and would thus form the output of Algorithm 2.1. However, f is not bounded from below on $\mathbb{R}$ and therefore does not possess a global minimal point. This unsolvability cannot be identified by Algorithm 2.1 and is therefore listed as an input requirement.

The second disadvantage of Algorithm 2.1 is that it must determine *all* critical points. However, with a nonlinear critical point equation $\nabla f(x) = 0$, it is rarely clear how all solutions can be determined. If critical points are overlooked, there is a risk that the output of Algorithm 2.1 is not a global minimal point.

A third disadvantage is that, for complicated functions f, even the calculation of a single critical point can be intricate.

At least Algorithm 2.1 can be applied to solvable optimization problems, whose critical points can be completely and explicitly calculated, for example by case distinction. Unfortunately, this often only applies to low-dimensional problems with a 'manageable' objective function.

Exercise 2.1.18 Consider again the data approximation problem from Example 1.1.4 with the Euclidean norm, i.e., the unconstrained optimization problem known as *linear regression*

$$
\min_{a,b} \left\| \begin{pmatrix} ax_1 + b - y_1 \\ \vdots \\ ax_m + b - y_m \end{pmatrix} \right\|_2
$$

for given data points $(x_j, y_j) \in \mathbb{R}^2$, $1 \le j \le m$. Why is it not guaranteed that Fermat's rule can be applied to every local minimal point of this problem?

Calculate all critical points of the equivalent optimization problem known as the *least squares problem* (Exercise 1.3.4)

$$
\min_{a,b} \left\| \begin{pmatrix} ax_1 + b - y_1 \\ \vdots \\ ax_m + b - y_m \end{pmatrix} \right\|_2^2
$$

under the condition that at least two of the points x_j, $1 \le j \le m$, are different from each other.

2.1.3 Geometric Properties of Gradients

To fully understand the geometric interpretation of the gradient $\nabla f(\bar{x})$, we associate it with the lower level set

$$
f_{\le}^{f(\bar{x})} = \{x \in \mathbb{R}^n \mid f(x) \le f(\bar{x})\}.
$$

This set is of fundamental importance for minimization methods, as it contains the points x which are 'at least as good' as $\bar{x}$ in terms of their objective value. The 'better' points x are just those that satisfy the strict inequality $f(x) < f(\bar{x})$. A

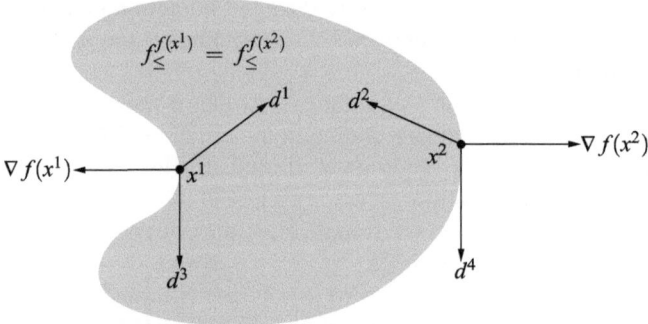

Fig. 2.2 Gradients and descent directions

descent direction d for f in $\bar{x}$ should therefore point into the 'interior' of $f_\leq^{f(\bar{x})}$. Figure 2.2 illustrates what such a set can look like for a nonlinear function f, with two different points x^1 and x^2 chosen so that $f_\leq^{f(x^1)} = f_\leq^{f(x^2)}$ holds.

Completely analogous to Lemma 2.1.10 and Definition 2.1.7, it can be shown that every vector $d \in \mathbb{R}^n$ with $\langle \nabla f(\bar{x}), d \rangle > 0$ is a first-order *ascent direction*. Since for a noncritical point $\bar{x}$ the gradient direction $d = \nabla f(\bar{x})$ satisfies the strict inequality

$$\langle \nabla f(\bar{x}), \nabla f(\bar{x}) \rangle = \|\nabla f(\bar{x})\|_2^2 > 0,$$

it is a first-order ascent direction for f at $\bar{x}$ and thus certainly points *out* of $f_\leq^{f(\bar{x})}$.

In fact, it can even be shown that $\nabla f(\bar{x})$ stands *perpendicular* to the boundary of $f_\leq^{f(\bar{x})}$. To justify this correctly, one must define a tangent cone to the set $f_\leq^{f(\bar{x})}$ at the point $\bar{x}$, which we postpone to Sect. 3.1.2, where tangent cones to general sets are introduced (Exercise 3.2.9).

For a preliminary argument, it is at least plausible that the tangent cone, i.e., the *linearization* of the set $f_\leq^{f(\bar{x})}$ around $\bar{x}$, is related to the linearization of the inequality that defines the set. Indeed, for $t > 0$ and a direction $d \in \mathbb{R}^n$, the point $\bar{x} + td$ lies in $f_\leq^{f(\bar{x})}$ if $f(\bar{x} + td) \leq f(\bar{x})$ holds. With the one-dimensional restriction, this can also be written as $\varphi_d(t) \leq \varphi_d(0)$. A linearization of this inequality around $\bar{t} = 0$, i.e., the first-order Taylor expansion with neglected remainder term (Theorem 2.1.19a), leads to

$$\varphi_d(0) + \varphi_d'(0)\,t \;\leq\; \varphi_d(0),$$

from which the condition $\langle \nabla f(\bar{x}), d \rangle = \varphi_d'(0) \leq 0$ follows for every $t > 0$. With the results from Sect. 3.2.2, we will actually be able to show in Exercise 3.2.9 that, at a noncritical point $\bar{x}$, the tangent cone to the set $f_\leq^{f(\bar{x})}$ at the point $\bar{x}$ is given by

$$\{d \in \mathbb{R}^n | \langle \nabla f(\bar{x}), d \rangle \leq 0\}.$$

Every direction d from the boundary of the tangent cone therefore fulfills $\langle \nabla f(\bar{x}), d \rangle = 0$, and is thus perpendicular to the gradient $\nabla f(\bar{x})$. Consequently, $\nabla f(\bar{x})$ not only points out of $f_{\leq}^{f(\bar{x})}$, but also stands perpendicular to the boundary of its tangent cone.

Figure 2.2 illustrates this for the points x^1 and x^2. Since d^1 and d^2 form obtuse angles with $\nabla f(x^1)$ and $\nabla f(x^2)$, respectively, they are first-order descent directions for f at x^1 and x^2. Along these directions one initially moves into the interior of the set $f_{\leq}^{f(x^1)} = f_{\leq}^{f(x^2)}$. The vector d^3 stands perpendicular to $\nabla f(x^1)$, so it is not a first-order descent direction. The fact that it still works as a descent direction is due to the 'nonconvex curvature' of the boundary of this set around x^1. On the other hand, the vector d^4 also stands perpendicular to $\nabla f(x^2)$, but does not qualify as a descent direction. We will deal with these two effects in more detail in Sect. 2.1.4.

While we have so far only considered the *sign* of the directional derivative $\varphi_d'(0) = f'(\bar{x}, d) = \langle \nabla f(\bar{x}), d \rangle$, we finally examine how to quantify the actual increase or decrease of function values from $\bar{x}$ in direction d. The problem here is that for every vector $d \neq 0$, the vector $2d$ points in the same direction, but when exchanging d for $2d$, the value of the directional derivative doubles. To obtain a unique and easy to interpret value of the directional derivative, it makes sense to consider only such directions d that have length one, i.e., satisfy $\|d\|_2 = 1$. This corresponds to the natural requirement that a step of length t from $\bar{x}$ in direction d leads to a point $\bar{x} + td$ that has the distance t from $\bar{x}$ (because, with $\|d\|_2 = 1$, $\|(\bar{x} + td) - \bar{x}\|_2 = t\|d\|_2 = t$ holds).

For such normalized directions d, the Cauchy-Schwarz inequality (which follows, e.g., from the representation (2.1) of the inner product) yields

$$-\|\nabla f(\bar{x})\|_2 = -\|\nabla f(\bar{x})\|_2 \cdot \|d\|_2 \leq \langle \nabla f(\bar{x}), d \rangle$$
$$\leq \|\nabla f(\bar{x})\|_2 \cdot \|d\|_2 = \|\nabla f(\bar{x})\|_2 \,,$$

and the lower and upper bounds are attained exactly for linearly dependent vectors d and $\nabla f(\bar{x})$. Due to $\nabla f(\bar{x}) \neq 0$ the smallest possible slope $-\|\nabla f(\bar{x})\|_2$ is therefore realized with

$$d = -\frac{\nabla f(\bar{x})}{\|\nabla f(\bar{x})\|_2}$$

and the largest possible slope $+\|\nabla f(\bar{x})\|_2$ with

$$d = +\frac{\nabla f(\bar{x})}{\|\nabla f(\bar{x})\|_2}.$$

In particular, the *length* $\|\nabla f(\bar{x})\|_2$ of the gradient exactly corresponds to the greatest possible increase of the function f from $\bar{x}$, and the *direction* of the gradient points in the corresponding direction of the *steepest ascent*.

Similarly, $-\nabla f(\bar{x})$ points in the direction of the *steepest descent* of f at $\bar{x}$. This will lead to a basic minimization algorithm in Sect. 2.2, the gradient method. In numerical algorithms, however, one does *not* work with normalized direction vectors, as for example the length $\|\nabla f(x)\|_2$ of the negative gradient direction is close to zero near the sought-after critical points, and the division $-\nabla f(x)/\|\nabla f(x)\|_2$ would then be numerically unstable.

2.1.4 Second-Order Optimality Conditions

To derive Fermat's rule, we exploited in Sect. 2.1.2 that there can be no *first-order* descent directions at local minimal points. However, Exercise 2.1.2 excludes *any* descent directions at local minimal points, and in Example 2.1.16 and with the direction d^3 in Fig. 2.2, we saw that there are other than first-order descent directions. The second-order optimality conditions derived in the current section are based on the concept of *second-order* descent directions. To introduce it, we assume f to be at least twice differentiable at the considered point $\bar{x} \in \mathbb{R}^n$ (without further mention). Then the one-dimensional restriction φ_d at $\bar{t} = 0$ is also twice differentiable. Its second derivative $\varphi_d''(0)$ is a measure for the curvature of φ_d at $\bar{t} = 0$. From Sect. 2.1.2 it is known that φ_d in the case $\varphi_d'(0) < 0$ initially decreases from $\bar{t} = 0$ for increasing values of t, and analogously that φ_d in the case $\varphi_d'(0) > 0$ initially increases from $\bar{t} = 0$ for increasing values of t. In the borderline case $\varphi_d'(0) = 0$ it seems plausible that for $\varphi_d''(0) < 0$ the function values of φ_d initially decrease with increasing t, so that such a d is a descent direction. To actually show this, we need Taylor's theorem (proven for example in [16, 18, 29]) in the following form, where the term *univariate* refers to the fact that the function under consideration depends on an only *one*-dimensional variable.

Theorem 2.1.19 (First and Second-Order Expansions by the Univariate Taylor Theorem)

(a) *Let $\varphi : \mathbb{R} \to \mathbb{R}$ be differentiable at $\bar{t}$. Then for all $t \in \mathbb{R}$*

$$\varphi(t) = \varphi(\bar{t}) + \varphi'(\bar{t})(t - \bar{t}) + o(|t - \bar{t}|)$$

holds, where $o(|t - \bar{t}|)$ denotes an expression of the form $\omega(t) \cdot |t - \bar{t}|$ with $\lim_{t \to \bar{t}} \omega(t) = \omega(\bar{t}) = 0$.

(b) *Let $\varphi : \mathbb{R} \to \mathbb{R}$ be twice differentiable at $\bar{t}$. Then for all $t \in \mathbb{R}$*

$$\varphi(t) = \varphi(\bar{t}) + \varphi'(\bar{t})(t - \bar{t}) + \tfrac{1}{2}\varphi''(\bar{t})(t - \bar{t})^2 + o(|t - \bar{t}|^2)$$

holds, where $o(|t - \bar{t}|^2)$ denotes an expression of the form $\omega(t) \cdot |t - \bar{t}|^2$ with $\lim_{t \to \bar{t}} \omega(t) = \omega(\bar{t}) = 0$.

Lemma 2.1.20 *For $f : \mathbb{R}^n \to \mathbb{R}$, a point $\bar{x} \in \mathbb{R}^n$ and a direction $d \in \mathbb{R}^n$, let $\varphi_d'(0) = 0$ and $\varphi_d''(0) < 0$ hold. Then d is a descent direction for f at $\bar{x}$.*

Proof As in the proof of Lemma 2.1.5, we assume that d is not a descent direction. Then for each $k \in \mathbb{N}$ there exists some $t^k \in (0, 1/k)$ with

$$\varphi_d(t^k) = f(\bar{x} + t^k d) \geq f(\bar{x}) = \varphi_d(0).$$

Moreover, Theorem 2.1.19b with $\varphi = \varphi_d$ and $\bar{t} = 0$ provides for every $k \in \mathbb{N}$

$$\varphi_d(t^k) = \varphi_d(0) + \varphi_d'(0)\, t^k + \tfrac{1}{2}\varphi_d''(0)(t^k)^2 + o((t^k)^2).$$

Together with the condition $\varphi_d'(0) = 0$ and $t^k \neq 0$, we obtain in total for every $k \in \mathbb{N}$

$$0 \leq \frac{\varphi_d(t^k) - \varphi_d(0)}{(t^k)^2} = \tfrac{1}{2}\varphi_d''(0) + \omega(t^k)$$

with $\lim_k \omega(t^k) = \omega(0) = 0$. Therefore, taking the limit we obtain

$$0 \leq \tfrac{1}{2}\varphi_d''(0)$$

which, however, contradicts the condition $\varphi_d''(0) < 0$. Hence, d is a descent direction. $\qquad\square$

Lemma 2.1.21 *Let $\bar{x}$ be a local minimal point of $f : \mathbb{R}^n \to \mathbb{R}$. Then $\nabla f(\bar{x}) = 0$ holds, and every direction $d \in \mathbb{R}^n$ satisfies $\varphi_d''(0) \geq 0$.*

Proof First, Theorem 2.1.13 provides that $\nabla f(\bar{x}) = 0$ holds and thus in particular $\varphi_d'(0) = \langle \nabla f(\bar{x}), d \rangle = 0$. From Exercise 2.1.2 and Lemma 2.1.20 the assertion follows. $\qquad\square$

In order to utilize Lemma 2.1.21, we need some formula for $\varphi_d''(0)$. It is the derivative of the first derivative

$$\varphi_d'(t) = \langle \nabla f(\bar{x} + td), d \rangle = \sum_{i=1}^{n} \partial_{x_i} f(\bar{x} + td)\, d_i$$

at the point $\bar{t} = 0$. From one more application of the chain rule, it follows

$$\varphi_d''(t) = \sum_{i=1}^{n} \left(D\partial_{x_i} f(\bar{x} + td) \, d \right) d_i = \sum_{i=1}^{n} \sum_{j=1}^{n} \partial_{x_j} \partial_{x_i} f(\bar{x} + td) \, d_j \, d_i$$

and thus

$$\varphi_d''(0) = \sum_{i=1}^{n} \sum_{j=1}^{n} \partial_{x_j} \partial_{x_i} f(\bar{x}) \, d_j \, d_i \, .$$

In this somewhat messy formula for $\varphi_d''(0)$, at least *partial* second derivatives of f appear. To find a more transparent formula for $\varphi_d''(0)$, we introduce an 'n-dimensional second derivative' of f. A straightforward way to do this is to form the first derivative of the first derivative, i.e., the Jacobian matrix of the gradient of f: The (n, n)-matrix

$$D^2 f(\bar{x}) := D\nabla f(\bar{x}) = \begin{pmatrix} \partial_{x_1} \partial_{x_1} f(\bar{x}) & \cdots & \partial_{x_n} \partial_{x_1} f(\bar{x}) \\ \vdots & & \vdots \\ \partial_{x_1} \partial_{x_n} f(\bar{x}) & \cdots & \partial_{x_n} \partial_{x_n} f(\bar{x}) \end{pmatrix}$$

is called the *Hessian matrix* of f at $\bar{x}$. As a second derivative, it encodes curvature information of f at $\bar{x}$.

According to the rules of matrix-vector multiplication, for every vector $d \in \mathbb{R}^n$ we have

$$d^{\mathsf{T}} D^2 f(\bar{x}) d = \sum_{i=1}^{n} \sum_{j=1}^{n} \partial_{x_j} \partial_{x_i} f(\bar{x}) \, d_j \, d_i = \varphi_d''(0),$$

which yields a short formula for $\varphi_d''(0)$. With this, we can reformulate Lemma 2.1.20.

Lemma 2.1.22 *For $f : \mathbb{R}^n \to \mathbb{R}$, a point $\bar{x} \in \mathbb{R}^n$ and a direction $d \in \mathbb{R}^n$, let $\langle \nabla f(\bar{x}), d \rangle = 0$ and $d^{\mathsf{T}} D^2 f(\bar{x}) d < 0$ hold. Then d is a descent direction for f at $\bar{x}$.*

This motivates the following definition.

Definition 2.1.23 (Second-Order Descent Direction) For $f : \mathbb{R}^n \to \mathbb{R}$ and $\bar{x} \in \mathbb{R}^n$, every $d \in \mathbb{R}^n$ with $\langle \nabla f(\bar{x}), d \rangle = 0$ and $d^{\mathsf{T}} D^2 f(\bar{x}) d < 0$ is called a *second-order descent direction* for f at $\bar{x}$.

Example 2.1.24 At $\bar{x} = 0$ the functions $f_1(x) = x_1^2 + x_2^2$, $f_2(x) = -x_1^2 - x_2^2$ and $f_3(x) = x_1^2 - x_2^2$ from Example 2.1.16 have the gradient $\nabla f(\bar{x}) = 0$ and the Hessian matrices

$$D^2 f_1(0) = \begin{pmatrix} 2 & 0 \\ 0 & 2 \end{pmatrix}, \qquad D^2 f_2(0) = \begin{pmatrix} -2 & 0 \\ 0 & -2 \end{pmatrix}, \qquad D^2 f_3(0) = \begin{pmatrix} 2 & 0 \\ 0 & -2 \end{pmatrix}.$$

For $d = (0, 1)^\mathsf{T}$ it follows

$$\langle \nabla f_2(0), d \rangle = \langle \nabla f_3(0), d \rangle = \langle 0, d \rangle = 0$$

and

$$d^\mathsf{T} D^2 f_2(0)d = d^\mathsf{T} D^2 f_3(0)d = -2 < 0,$$

so that d is a second-order descent direction for f_2 and f_3 at $\bar{x} = 0$.

Example 2.1.25 In Example 2.1.24, the condition $\langle \nabla f(\bar{x}), d \rangle = 0$ from Definition 2.1.23 is fulfilled, because even $\nabla f(\bar{x}) = 0$ holds. However, there are also second-order descent directions in the case $\nabla f(\bar{x}) \neq 0$, namely orthogonal ones to $\nabla f(\bar{x})$. This is illustrated by the direction d^3 in Fig. 2.2. The direction d^4 there, on the other hand, makes it clear that mere orthogonality is not sufficient. The 'nonconvex curvature' of the boundary of the set $f_{\leq}^{f(x^1)} = f_{\leq}^{f(x^2)}$ at x^1 corresponds exactly to the condition $(d^3)^\mathsf{T} D^2 f(x^1)d^3 < 0$, which according to Definition 2.1.23 makes d^3 a second-order descent direction. This connection will become even clearer when we link these conditions with the eigenvalues of the matrix $D^2 f(x^1)$.

The following example shows that not every descent direction is either of first or second order.

Example 2.1.26 The function $f(x) = x^3$ satisfies $f'(\bar{x}) = 0$ at $\bar{x} = 0$, so that its descent direction $d = -1$ at $\bar{x}$ is not of first order. Moreover, it is not hard to see that, for every function f and for every second-order descent direction d at some point $\bar{x}$, its 'opposite direction' $-d$ is also a second-order descent direction at $\bar{x}$. Since in the example function the opposite direction $d = 1$ is an ascent direction for f at $\bar{x} = 0$, $d = -1$ is neither a second-order descent direction.

The above formula for $\varphi_d''(0)$ provides a more explicit formulation of Lemma 2.1.21, stating that at a local minimal point $\bar{x}$ of f necessarily $\nabla f(\bar{x}) = 0$ and $d^\mathsf{T} D^2 f(\bar{x})d \geq 0$ hold for all $d \in \mathbb{R}^n$. In linear algebra, the latter condition on the matrix $D^2 f(\bar{x})$ is called *positive semidefiniteness* and is abbreviated as $D^2 f(\bar{x}) \succeq 0$. Thus the reformulation of Lemma 2.1.21 takes the following form.

Theorem 2.1.27 (Second-Order Necessary Optimality Condition) *Let the function $f : \mathbb{R}^n \to \mathbb{R}$ be twice differentiable at a local minimal point $\bar{x} \in \mathbb{R}^n$. Then $\nabla f(\bar{x}) = 0$ and $D^2 f(\bar{x}) \succeq 0$ hold.*

In order to apply Theorem 2.1.27 algorithmically, the condition $D^2 f(\bar{x}) \succeq 0$ must be verifiable in finitely many steps. Checking positive semidefiniteness using its definition, however, would require to verify infinitely many inequalities. Fortunately, linear algebra provides a characterization of positive semidefiniteness, provided the matrix $D^2 f(\bar{x})$ is *symmetric* (i.e., $D^2 f(\bar{x}) = D^2 f(\bar{x})^\mathsf{T}$ holds). According to Schwarz's theorem (e.g., [17]), the latter is the case if f is not only twice differentiable, but even twice *continuously* differentiable (shortly: $f \in C^2(\mathbb{R}^n, \mathbb{R})$).

Recall that λ is an *eigenvalue* to the *eigenvector* $v \neq 0$ of $D^2 f(\bar{x})$ if $D^2 f(\bar{x})v = \lambda v$ holds (a motivation of this property is given, for example, in the appendix of [25]). While eigenvalues can generally be complex numbers, it is shown in linear algebra (e.g., [8, 20]) that eigenvalues of *symmetric* matrices are always real. In particular, one can then consider their signs. A symmetric matrix is indeed positive semidefinite if and only if all its eigenvalues are nonnegative (e.g., [8, 20]). Therefore, we may verify the condition $D^2 f(\bar{x}) \succeq 0$ for any C^2-function f by calculating the n eigenvalues of the matrix $D^2 f(\bar{x})$ and checking them for nonnegativity.

Example 2.1.28 All three functions f_1, f_2 and f_3 from Example 2.1.24 are twice continuously differentiable at $\bar{x} = 0$, so the positive semidefiniteness of their Hessian matrices can be checked using the eigenvalues. Of the three Hessian matrices, only $D^2 f_1(0)$ is positive semidefinite. Therefore, using Theorem 2.1.27, we can rule out that f_2 and f_3 have a local minimal point at $\bar{x} = 0$.

Example 2.1.28 shows that Theorem 2.1.27 can significantly reduce the set of candidates for local minimal points compared to Fermat's rule. This forms the basis of Algorithm 2.2, which uses a correspondingly 'finer filter' compared to Algorithm 2.1. The three main disadvantages of Algorithm 2.1, namely its inability to identify the unsolvability of the optimization problem, the need to completely calculate the set of candidates C, and the difficulty of determining critical points in the first place, prevail in Algorithm 2.2. Its advantage over Algorithm 2.1 is the smaller set C.

Unfortunately, local minimal points are still not *characterized* by the necessary condition from Theorem 2.1.27 because, e.g., for $f_4(x) = x_1^2 - x_2^4$ we have $\nabla f_4(0) = 0$, the Hessian matrix

$$D^2 f_4(0) = \begin{pmatrix} 2 & 0 \\ 0 & 0 \end{pmatrix}$$

Algorithm 2.2: Conceptual algorithm for unconstrained nonlinear minimization with second-order information

Input: Solvable unconstrained twice continuously differentiable optimization problem P
Output: Global minimal point $x^\star$ of f over $\mathbb{R}^n$

1 **begin**
2   Determine all critical points with positive semidefinite Hessian matrix of f, i.e., the solution set C of the two conditions $\nabla f(x) = 0$ and $D^2 f(x) \succeq 0$.
3   Determine a minimal point $x^\star$ of f in C.
4 **end**

---

is positive semidefinite, but $\bar{x} = 0$ is not a local minimal point of f. This motivates the need for a *sufficient* condition for local minimality.

Analogous to our previous considerations, one might suspect that f certainly has a local minimal point at $\bar{x}$ if for all $d \in \mathbb{R}^n$ the one-dimensional restriction φ_d has a local minimal point at $\bar{t} = 0$. The following exercise, however, shows that this conjecture is *wrong*.

Exercise 2.1.29 (Example of Peano)   Show that the function $f(x) = (x_1^2 - x_2) \cdot (x_1^2 - 3x_2)$ does not have a local minimal point at $\bar{x} = 0$, but that for every direction $d \in \mathbb{R}^n$ the one-dimensional restriction φ_d has a local minimal point at $\bar{t} = 0$.

Exercise 2.1.29 illustrates an effect that cannot occur for univariate functions (i.e., for $n = 1$): The function value at $\bar{x}$ is diminished by function values at points that do not lie along a straight line through $\bar{x}$, but along a *parabola* through $\bar{x}$. In Exercise 2.1.29, for the direction d being tangential to this parabola at $\bar{x}$, $\varphi_d''(0) = 0$ holds, while all other directions satisfy $\varphi_d''(0) > 0$.

To guarantee a local minimal point $\bar{x}$ through second-order information, it may thus be helpful to preclude in the second-order necessary condition $\nabla f(\bar{x}) = 0$, $D^2 f(\bar{x}) \succeq 0$, that some direction d satisfies only $\varphi_d''(0) = 0$. In view of $\langle \nabla f(\bar{x}), d \rangle = 0$ this is equivalent to the requirement that every direction d is a second-order *ascent* direction, i.e.,

$$d^\mathsf{T} D^2 f(\bar{x}) d > 0 \quad \text{for all } d \neq 0,$$

which in linear algebra is referred to as *positive definiteness* of $D^2 f(\bar{x})$, shortly $D^2 f(\bar{x}) \succ 0$. If the Hessian matrix is symmetric at $\bar{x}$ due to twice continuous differentiability of f, positive definiteness is characterized by all eigenvalues of $D^2 f(\bar{x})$ being strictly positive (e.g., [8, 20]).

The proof of the resulting sufficient second-order optimality condition again uses Taylor's theorem. Exercise 2.1.29 suggests, however, that univariate versions of this theorem, i.e., expansions along straight lines, will not be helpful.

Fortunately, Taylor's theorem can also be formulated in the multivariate setting (proofs are given, for example, in [16, 17]).

Theorem 2.1.30 (First and Second-Order Expansions by the Multivariate Taylor Theorem)

(a) *Let* $f : \mathbb{R}^n \to \mathbb{R}$ *be differentiable at* $\bar{x}$. *Then for all* $x \in \mathbb{R}^n$

$$f(x) = f(\bar{x}) + \langle \nabla f(\bar{x}), x - \bar{x} \rangle + o(\|x - \bar{x}\|)$$

holds, where $o(\|x - \bar{x}\|)$ *denotes an expression of the form* $\omega(x) \cdot \|x - \bar{x}\|$ *with* $\lim_{x \to \bar{x}} \omega(x) = \omega(\bar{x}) = 0$.

(b) *Let* $f : \mathbb{R}^n \to \mathbb{R}$ *be twice differentiable at* $\bar{x}$. *Then for all* $x \in \mathbb{R}^n$

$$f(x) = f(\bar{x}) + \langle \nabla f(\bar{x}), x - \bar{x} \rangle + \tfrac{1}{2}(x - \bar{x})^{\mathsf{T}} D^2 f(\bar{x})(x - \bar{x}) + o(\|x - \bar{x}\|^2)$$

holds, where $o(\|x - \bar{x}\|^2)$ *denotes an expression of the form* $\omega(x) \cdot \|x - \bar{x}\|^2$ *with* $\lim_{x \to \bar{x}} \omega(x) = \omega(\bar{x}) = 0$.

In the proof of the following theorem, we will denote a ball with radius r around $\bar{x}$ with

$$B_{\leq}(\bar{x}, r) = \{x \in \mathbb{R}^n \mid \|x - \bar{x}\| \leq r\}$$

and its boundary, i.e., the sphere with radius r around $\bar{x}$, with

$$B_{=}(\bar{x}, r) = \{x \in \mathbb{R}^n \mid \|x - \bar{x}\| = r\}$$

(where the choice of the norm $\| \cdot \|$ does not matter).

Theorem 2.1.31 (Second-Order Sufficient Optimality Condition) *Let the function* $f : \mathbb{R}^n \to \mathbb{R}$ *be twice differentiable at* $\bar{x} \in \mathbb{R}^n$ *and let* $\nabla f(\bar{x}) = 0$ *and* $D^2 f(\bar{x}) \succ 0$ *hold. Then* $\bar{x}$ *is a strict local minimal point of* f.

Proof The proof is provided by contradiction. Suppose, $\bar{x}$ is not a strict local minimal point of f. If $\bar{x}$ is not a critical point of f, a contradiction already occurs, so we can assume $\nabla f(\bar{x}) = 0$ in the following. Since $\bar{x}$ is not a strict local minimal point, by Definition 1.1.2 for every neighborhood U of $\bar{x}$ there exists some $x_U \in U \setminus \{\bar{x}\}$ with $f(x_U) \leq f(\bar{x})$. In particular, for every neighborhood $U_k = B_\leq(\bar{x}, 1/k)$ with $k \in \mathbb{N}$ there exists some point $x^k \neq \bar{x}$ with $f(x^k) \leq f(\bar{x})$. From the specific choice of the neighborhoods follows $\lim_k x^k = \bar{x}$.

Consequently, the values $t^k := \|x^k - \bar{x}\|$ form a null sequence of positive numbers, the directions

$$d^k := \frac{x^k - \bar{x}}{t^k} = \frac{x^k - \bar{x}}{\|x^k - \bar{x}\|}$$

are well-defined and normalized, and $x^k = \bar{x} + t^k d^k$ holds for all $k \in \mathbb{N}$. That all directions d^k are normalized means that the sequence (d^k) lies in the unit sphere $B_=(0, 1)$. Since the latter is compact, (d^k) has at least one accumulation point $d \in B_=(0, 1)$ (according to the Bolzano-Weierstrass theorem; e.g. [17]). After transitioning to a corresponding subsequence, we obtain the existence of sequences (t^k) and (d^k) with $\lim_k t^k = 0$, $\lim_k d^k = d$, $\|d\| = 1$ and

$$f(x^k) = f(\bar{x} + t^k d^k) \leq f(\bar{x})$$

for all $k \in \mathbb{N}$. Theorem 2.1.30b thus provides for all $k \in \mathbb{N}$

$$0 \geq \frac{f(\bar{x} + t^k d^k) - f(\bar{x})}{(t^k)^2} = \tfrac{1}{2}(d^k)^\mathsf{T} D^2 f(\bar{x}) d^k + \omega(x^k),$$

where $\omega(x^k)$ tends to $\omega(\bar{x}) = 0$ for $k \to \infty$. Taking the limit results in

$$0 \geq d^\mathsf{T} D^2 f(\bar{x}) d.$$

However, due to $\|d\| = 1$, this contradicts the condition $D^2 f(\bar{x}) \succ 0$. □

Remark 2.1.32 The *necessary* first and second-order optimality conditions and their proofs can easily be transferred to functionals on Banach spaces $f : X \to \mathbb{R}$, i.e., to a large class of infinite-dimensional optimization problems. The generalization of *sufficient* conditions, on the other hand, is only possible with additional effort. A main reason for this is that the unit sphere in a Banach space is only compact in special cases (for details see e.g. [4, 19]).

Exercise 2.1.33 Show that the unique critical point of the least squares problem

$$
\min_{a,b} \left\| \begin{pmatrix} ax_1 + b - y_1 \\ \vdots \\ ax_m + b - y_m \end{pmatrix} \right\|_2^2
$$

calculated in Exercise 2.1.18 under the condition that at least two of the points x_j, $1 \le j \le m$, are different, is a strict local minimal point.

Remark 2.1.34 One might wonder why there are a necessary and a sufficient version of the second-order optimality conditions, but only a necessary condition of first order, Fermat's rule. To answer this question, recall that the proof of Fermat's rule is based on Lemma 2.1.6 for one-sided directionally differentiable functions, i.e., on the necessity of stationarity for every local minimal point. In the smooth case, this means that a local minimal point $\bar{x}$ of f necessarily satisfies the inequalities $\langle \nabla f(\bar{x}), d \rangle \ge 0$ for all $d \in \mathbb{R}^n$.

If one were to replace the nonstrict with strict inequalities for the construction of a possibly sufficient first-order optimality condition, analogous to the second-order conditions, one would obtain $\langle \nabla f(\bar{x}), d \rangle > 0$ for all $d \in \mathbb{R}^n \setminus \{0\}$. However, this condition cannot be fulfilled for any vector $\nabla f(\bar{x})$ and is therefore useless.

The problem lies in the smoothness requirement for the function f. For functions that are only one-sided directionally differentiable, it can indeed be shown that from $f'(\bar{x}, d) > 0$ for all $d \in \mathbb{R}^n \setminus \{0\}$, the (strict) local minimality of $\bar{x}$ for f follows and that this condition can also be fulfilled (e.g., for $f(x) = |x|$ and $\bar{x} = 0$).

Also for *constrained* smooth problems we will be able to formulate a sufficient first-order optimality condition (Corollary 3.2.69) because, loosely speaking, the required nonsmoothness is provided by the geometry of the boundary of the feasible set.

Since the sufficient condition from Theorem 2.1.31 provides a bit more information than desired, namely even *strict* local minimal points, one cannot expect a *characterization* of local minimality from it. On the other hand, this sufficient condition neither provides a characterization for strict local minimality, as it can be violated at strict local minimal points (e.g., for $f(x) = x^4$ and $\bar{x} = 0$).

In this sense, there is a gap between necessary and sufficient second-order optimality conditions. However, the following results show that this gap does not matter for 'almost all' optimization problems. As a consequence, we do not deal with optimality conditions or descent directions of third and higher order, although these could be specified using the corresponding Taylor expansions. For the following definition, remember that a square matrix is *nonsingular* if none of its eigenvalues is zero.

Definition 2.1.35 (Nondegenerate Critical and Minimal Points) Let the function $f : \mathbb{R}^n \to \mathbb{R}$ be twice differentiable at $\bar{x}$ with $\nabla f(\bar{x}) = 0$. Then $\bar{x}$ is called

(a) *nondegenerate critical point*, if $D^2 f(\bar{x})$ is nonsingular,
(b) *nondegenerate local minimal point*, if $\bar{x}$ is a local minimal point and a nondegenerate critical point.

For example, the saddle point $\bar{x} = 0$ of $f_3(x) = x_1^2 - x_2^2$ is a nondegenerate critical point, while the saddle point $\bar{x} = 0$ of $f_4(x) = x_1^2 - x_2^4$ is a *degenerate* critical point.

Nondegenerate local minimal points can be *characterized* by properties of the gradient and Hessian matrix.

Lemma 2.1.36 *The point $\bar{x}$ is a nondegenerate local minimal point of f if and only if $\nabla f(\bar{x}) = 0$ and $D^2 f(\bar{x}) \succ 0$ hold.*

Proof For a local minimal point $\bar{x}$, according to Theorem 2.1.27, $\nabla f(\bar{x}) = 0$ and $D^2 f(\bar{x}) \succeq 0$ are satisfied. If $\bar{x}$ is also a nondegenerate critical point, then $D^2 f(\bar{x})$ has no vanishing eigenvalue, so it is positive definite.

On the other hand, let $\nabla f(\bar{x}) = 0$ and $D^2 f(\bar{x}) \succ 0$. Then, according to Theorem 2.1.31, $\bar{x}$ is a local minimal point. Since a positive definite matrix is nonsingular, this local minimal point is also nondegenerate. □

In a certain sense, which we will only briefly motivate, 'almost all' C^2-functions possess exclusively nondegenerate critical points. More precisely, the subset

$$\mathscr{F} = \{ f \in C^2(\mathbb{R}^n, \mathbb{R}) |\ \text{all critical points of } f \text{ are nondegenerate} \}$$

of the set of all C^2-functions is 'very large'. In fact, it is easy to see the equivalence

$$f \in \mathscr{F} \quad \Leftrightarrow \quad \forall x \in \mathbb{R}^n : \quad \|\nabla f(x)\| + |\det(D^2 f(x))| > 0,$$

so that $\mathscr{F}$ is an *open* set in a suitably chosen topology on the function space $C^2(\mathbb{R}^n, \mathbb{R})$ (which takes into account the first and second derivatives of a function for the definition of a neighborhood). If the strong Whitney topology (aka C_s^2-topology) is used for this, even the following result applies, for whose proof we refer to [22].

Theorem 2.1.37 *The set $\mathscr{F}$ is C_s^2-open and -dense in $C^2(\mathbb{R}^n, \mathbb{R})$.*

The openness part of Theorem 2.1.37 means that the elements of $\mathscr{F}$ are stable in the sense that they remain in $\mathscr{F}$ under sufficiently small perturbations. The denseness part states that elements from outside of $\mathscr{F}$ can be approximated arbitrarily well by elements of $\mathscr{F}$. In this sense, the nondegeneracy of a critical point and in particular the nondegeneracy of a local minimal point are therefore *mild* requirements.

Finally, we note that, analogous to the different interpretations of the sign and the value of the inner product $\langle \nabla f(\bar{x}), d \rangle$, not only the *signs* of the eigenvalues of $D^2 f(\bar{x})$ contain important information, but also their actual values. Indeed, let λ be an eigenvalue of $D^2 f(\bar{x})$, and let d with $\|d\|_2 = 1$ be a corresponding eigenvector. Then

$$\varphi_d''(0) \;=\; d^\mathsf{T} D^2 f(\bar{x}) d \;=\; d^\mathsf{T}(\lambda d) \;=\; \lambda \|d\|_2^2 \;=\; \lambda$$

holds, so that the size of the eigenvalue λ can be interpreted as a measure of the curvature of the one-dimensional restriction φ_d at $\bar{t} = 0$. Furthermore, for symmetric (n, n)-matrices it is known from linear algebra that the n eigenvectors of the n eigenvalues can be chosen to be pairwise orthogonal to each other (e.g. [8, 20]). This allows to describe the local structure of f around a nondegenerate critical point quite well.

Ultimately, these relationships also make it possible to explicitly specify second-order descent directions at critical points. For this purpose, let a critical point $\bar{x}$ of f with $D^2 f(\bar{x}) \nsucceq 0$ be given, i.e., $D^2 f(\bar{x})$ has (at least) one negative eigenvalue λ. Each corresponding eigenvector d is then a second-order descent direction, because it satisfies $\langle \nabla f(\bar{x}), d \rangle = 0$ and

$$d^\mathsf{T} D^2 f(\bar{x}) d \;=\; \lambda \;<\; 0.$$

Exercise 2.1.38 Show that at a nondegenerate saddle point, both a descent and an ascent direction of second order exist.

2.1.5 Convex Optimization Problems

Section 2.1 has provided optimality conditions for *local*, but not necessarily *global* minimal points of a function using first and second derivatives. Since derivatives only contain local information about a function, this is all one may expect, unless the function additionally possesses some appropriate *global property*. Such a property is convexity. Since convex optimization problems are extensively discussed in [36], we only summarize some essential results below and refer to [36] for proofs and further considerations.

> **Definition 2.1.39 (Convex Sets and Functions)**
>
> (a) A set $X \subseteq \mathbb{R}^n$ is called *convex*, if
>
> $$\forall x, y \in X, \ \lambda \in (0, 1) : \quad (1 - \lambda)x + \lambda y \in X$$
>
>   holds (i.e., the connecting line between each pair of two points in X belongs entirely to X; Fig. 2.3).
> (b) For a convex set $X \subseteq \mathbb{R}^n$ a function $f : X \to \mathbb{R}$ is called *convex (on X)*, if
>
> $$\forall x, y \in X, \ \lambda \in (0, 1) : \quad f((1 - \lambda)x + \lambda y) \leq (1 - \lambda)f(x) + \lambda f(y)$$
>
>   holds (i.e., the graph of f lies *below* each of its secants; Fig. 2.4).

While the convexity of a function is geometrically defined by the fact that its graph lies *below* each of its secants, the convexity of a continuously differentiable function f can be characterized by the fact that its graph lies *above* the graph of

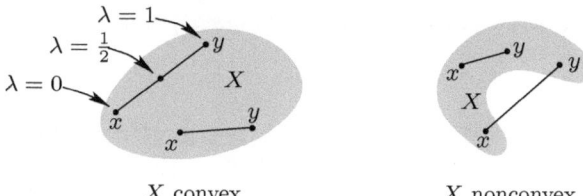

Fig. 2.3 Convexity of sets in $\mathbb{R}^2$

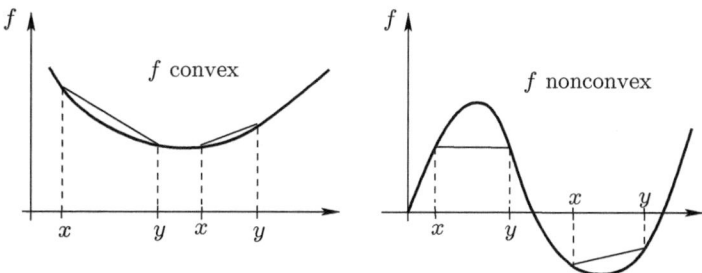

Fig. 2.4 Convexity of functions on $\mathbb{R}$

each of its linearizations. To formulate this result, for f we consider the multivariate Taylor expansion of first order from Theorem 2.1.30a around a point x, which states that for all $y \in \mathbb{R}^n$ the representation

$$f(y) = f(x) + \langle \nabla f(x), y - x \rangle + o(\|y - x\|)$$

applies. For linearization, we omit the error term $o(\|y - x\|)$ and obtain the (in y linear) function $f(x) + \langle \nabla f(x), y - x \rangle$ as an approximation of f at the point x. The announced characterization of convexity of continuously differentiable functions is therefore as follows.

> **Theorem 2.1.40 (C^1-Characterization of Convexity)** *On a convex set $X \subseteq \mathbb{R}^n$, a function $f \in C^1(X, \mathbb{R})$ is convex if and only if*
>
> $$\forall \, x, y \in X : \quad f(y) \geq f(x) + \langle \nabla f(x), y - x \rangle$$
>
> *applies.*

The central theorem for continuously differentiable convex unconstrained optimization problems is the following far-reaching extension of Fermat's rule.

> **Corollary 2.1.41** *Let the function $f \in C^1(\mathbb{R}^n, \mathbb{R})$ be convex. Then the critical points of f are exactly the global minimal points of f.*

Proof That every global minimal point is a critical point of f, follows from Theorem 2.1.13. On the other hand, according to Theorem 2.1.40 every point $\bar{x}$ with $\nabla f(\bar{x}) = 0$ satisfies

$$\forall \, y \in \mathbb{R}^n : \quad f(y) \geq f(\bar{x}) + \langle \underbrace{\nabla f(\bar{x})}_{=\,0}, y - \bar{x} \rangle = f(\bar{x})$$

and is thus a global minimal point. $\square$

The problem of global minimization of continuously differentiable convex functions is thus equivalent to a root-finding problem, namely to the solution of the equation $\nabla f(x) = 0$.

Although necessary or sufficient optimality conditions of *second* order are apparently superfluous in the convex case, Hessian matrices still play an important role, namely to provide a simple criterion for *proving* the convexity of twice continuously differentiable functions.

> **Theorem 2.1.42 (C^2-Characterization of Convexity)** *A function $f \in C^2(\mathbb{R}^n, \mathbb{R})$ is convex if and only if*
>
> $$\forall\, x \in \mathbb{R}^n : \quad D^2 f(x) \succeq 0$$
>
> *holds.*

To conclude this section, we consider a nondegenerate local minimal point $\bar{x}$ of $f \in C^2(\mathbb{R}^n, \mathbb{R})$, i.e., one with $D^2 f(\bar{x}) \succ 0$. Due to the continuity of $D^2 f$ and the continuous dependence of the eigenvalues of symmetric matrices on the matrix entries [37], $D^2 f(x) \succ 0$ also holds for points x from an entire neighborhood of $\bar{x}$. According to Theorem 2.1.42, f is therefore locally convex around the nondegenerate local minimal point $\bar{x}$.

Exercise 2.1.43 Consider the quadratic function $q(x) = \frac{1}{2}x^{\mathsf{T}} A x + b^{\mathsf{T}} x$ with $A = A^{\mathsf{T}} \succ 0$ and $b \in \mathbb{R}^n$. Show that q is a convex function on $\mathbb{R}^n$ and that its unique minimal point is

$$x^{\star} = -A^{-1} b$$

with minimal value

$$q(x^{\star}) = -\tfrac{1}{2}\, b^{\mathsf{T}} A^{-1} b.$$

We remark that the function q from Exercise 2.1.43 is even *strongly convex* [36].

Exercise 2.1.44 Show that the objective function of the optimization problem

$$\min_{a,b} \left\| \begin{pmatrix} ax_1 + b - y_1 \\ \vdots \\ ax_m + b - y_m \end{pmatrix} \right\|_2^2$$

from Exercise 2.1.18 is convex, and determine under the assumption that at least two of the points x_j, $1 \le j \le m$, are different from each other, the unique global minimal point as well as the global minimal value.

2.2   Algorithms

In this section, we develop algorithms for minimizing a smooth function $f :$ $\mathbb{R}^n \to \mathbb{R}$, where the vague term 'smooth' means that the respective continuity and differentiability requirements are met. All presented methods start from a user-provided starting point x^0 and generate iteratively a sequence (x^k), whose accumulation points are at least critical points of f, i.e., zeros of the gradient ∇f. We will see that this sequence converges under certain conditions, and that its limit point will usually be a local minimal point of f. One should *not* expect to find a *global* minimal point of f algorithmically in this way, unless additional global information about f is available. For methods of global optimization, we instead refer to [36].

Section 2.2.1 initially introduces a so-called descent method as a very general framework, without explicitly shaping it yet. Nevertheless, it is possible to formulate sufficient conditions for the termination of this procedure, which we can later check for the explicitly stated procedures. The new iterates are generated by a combination of search direction vectors and step sizes along these search directions, and the aforementioned sufficient conditions are the so-called gradient-relatedness of the search direction sequence and the efficiency of the step size sequence. Section 2.2.2 presents three possibilities for determining efficient step size sequences, which can then be used in the concrete descent methods.

In Sect. 2.2.3 we examine the most obvious choice for constructing gradient-related search direction sequences, which leads to the gradient method. However, it proves to be very slow in practice, which we quantify by introducing various convergence speeds. With the geometric insight that the main reason for the slowness of the gradient method lies in the lack of utilization of curvature information of the objective function, we modify it in Sect. 2.2.4 to the class of variable metric methods.

An important representative of this class of methods is the Newton method discussed in Sect. 2.2.5. Although its quadratic convergence speed is very fast, it has the decisive disadvantage of being a descent method only under often unrealistic conditions. We therefore modify our approach further and in Sect. 2.2.6 state general conditions under which variable metric methods at least converge superlinearly, before Sect. 2.2.7 introduces the quasi-Newton methods based on this.

The fact that quasi-Newton methods rely on curvature information from approximations of the Hessian matrix of the objective function can lead to storage space problems in high-dimensional optimization problems. Therefore, we try to improve the gradient method also by matrix-free methods. Surprisingly, a clever combination of gradient information can gain so much curvature information that such methods actually exist. These are the conjugate gradient methods, based on the concept of conjugate directions explained in Sect. 2.2.8 and introduced in Sect. 2.2.9.

Section 2.2.10 finally deals with trust-region methods, which, unlike the other discussed descent methods, do not first determine a search direction and then a step size, but first a search radius and then the direction to the new iterate.

2.2.1 Descent Methods

In addition to the continuity of the objective function f, we require throughout Sect. 2.2 that the lower level set $f_{\leq}^{f(x^0)}$ for the starting point $x^0 \in \mathbb{R}^n$ is bounded. If this assumption is violated, the presented convergence proofs cannot be carried out, and the considered methods can then only be regarded as heuristics. According to Lemma 1.2.26, however, the requirement is fulfilled for every starting point x^0 if f is coercive on $\mathbb{R}^n$.

As the first reason for the introduction of this assumption, we note that for bounded $f_{\leq}^{f(x^0)}$ a continuous function f has a global minimal point according to Corollary 1.2.17 and that then the equation $\nabla f(x) = 0$ is solvable in the first place.

Remark 2.2.1 For a bounded set $f_{\leq}^{f(x^0)}$, the function f is therefore bounded from below. In the literature, the latter weaker assumption is sometimes used for convergence proofs. However, the subsequent assumption of Lipschitz continuity of the gradient ∇f on $f_{\leq}^{f(x^0)}$ becomes a strong or even unsatisfiable requirement in many applications, if the set $f_{\leq}^{f(x^0)}$ is allowed to be unbounded (Remark 2.2.11).

As a first algorithmic idea, one could try to solve the equation $\nabla f(x) = 0$ with Newton's method known from numerical analysis

$$x^{k+1} = x^k - (D^2 f(x^k))^{-1} \nabla f(x^k), \quad k = 0, 1, 2, \ldots$$

(for a geometric interpretation see [25, 36]). The advantage would be a high convergence speed if x^0 is close enough to a solution. The disadvantages are that x^0 does not need to be close to a solution, that the Hessian matrix $D^2 f(x^k)$ does not necessarily have to be invertible (and possibly difficult to evaluate) and that the Newton method can also converge to local maximal points and saddle points.

Although, due to its high local convergence speed, we will return in detail to the idea of the Newton method in Sect. 2.2.5, we first consider methods that generate a *descent* in the objective function value at each iteration step, i.e., they fulfill

$$\forall\, k \in \mathbb{N}_0: \quad f(x^{k+1}) < f(x^k).$$

Such methods may only converge to local maximal points 'under very unfortunate circumstances' (e.g., if x^0 is a local maximal point or if a long descent step skips a local minimal point and 'accidentally' hits a local maximal point), and from geometric considerations, convergence to saddle points is unlikely (x^0 would have to lie in a lower-dimensional manifold for a descent method to converge to a saddle point, which is not to be expected in practical algorithms due to ubiquitous small numerical perturbations).

A general descent method is formulated in Algorithm 2.3. In its input and subsequently, we speak somewhat loosely of a 'C^1-optimization problem P' when the defining functions of P are continuously differentiable.

From a theoretical point of view, differentiability would often suffice. However, in applications, differentiable functions are typically also continuously differentiable, so this requirement is common and does not pose a significant limitation.

Algorithm 2.3: General descent method

Input: C^1-optimization problem P, starting point x^0 and termination tolerance $\varepsilon > 0$
Output: Approximation $\bar{x}$ of a critical point of f (if the method terminates; Corollary 2.2.10)

1 **begin**
2   Set $k = 0$.
3   **while** $\|\nabla f(x^k)\| > \varepsilon$ **do**
4   Choose x^{k+1} with $f(x^{k+1}) < f(x^k)$.
5   Replace k with $k + 1$.
6   **end**
7   Set $\bar{x} = x^k$.
8 **end**

Different descent methods differ by the choice of x^{k+1} in line 4 of Algorithm 2.3. In the following, we will first derive mild conditions on the choice of x^{k+1}, independent of the specific design of line 4, which guarantee that Algorithm 2.3 actually terminates after a finite number of steps.

In the case when these mild conditions are violated, Algorithm 2.3 is often equipped with an 'emergency brake', namely with the additional termination criterion $k > k_{\max}$ with a high iteration number $k_{\max}$ (such as $k_{\max} = 10^4 \cdot n$). One can then not expect that the output $\bar{x}$ lies close to a critical point of f, but at least it fulfills the inequality $f(\bar{x}) < f(x^0)$. For clarity, we will omit the explicit consideration of the 'emergency brake' in the following.

Remark 2.2.2 In behavioral economics and psychology, the concept of *bounded rationality* explains that decision-makers, faced with the task of solving an optimization problem, are sometimes satisfied with finding any point $\bar{x}$ with a better objective function value than x^0, i.e., with $f(\bar{x}) < f(x^0)$. Apparently, a finite number of steps of any descent method (as well as of improvement heuristics [25]) are sufficient for this, without the termination criterion in line 3 needing to be fulfilled. The implementation of the above-described 'emergency brake' guarantees such an output $\bar{x}$ for each descent method.

In line 3 of Algorithm 2.3, one tests $\|\nabla f(x^k)\| > \varepsilon$ with a tolerance $\varepsilon > 0$, as one cannot expect to algorithmically determine a critical point *exactly* (a while-loop with the condition $\|\nabla f(x^k)\| > 0$ would in most cases never terminate). The generated output $\bar{x}$ with $\|\nabla f(\bar{x})\| \leq \varepsilon$ is then only an *approximation* of a critical point.

To guarantee that Algorithm 2.3 terminates after a finite number of iterations, one must prove that independent of the choice of ε, some $k \in \mathbb{N}$ with $\|\nabla f(x^k)\| \leq \varepsilon$ exists. This is certainly guaranteed if the sequence $(\nabla f(x^k))$ converges to the zero vector. However, it is also sufficient, for example, that this sequence merely has the zero vector as an accumulation point. Since, on the other hand, for a given $\varepsilon > 0$ in the case of termination of Algorithm 2.3, no infinite sequence would be generated, we will artificially set $\varepsilon = 0$ for the following convergence investigations and infer from the obtained results to the finite termination in the case $\varepsilon > 0$.

First, we investigate whether the iterates x^k themselves and their function values converge.

Lemma 2.2.3 *For bounded $f_{\leq}^{f(x^0)}$ the sequence (x^k) generated by Algorithm 2.3 with $\varepsilon = 0$ either terminates after finitely many steps with a critical point, or it possesses at least one accumulation point in $f_{\leq}^{f(x^0)}$, and the sequence of function values $(f(x^k))$ converges.*

Proof Due to the descent property in line 4, all iterates x^k lie in the set $f_{\leq}^{f(x^0)}$. The latter is bounded by assumption and also closed (Exercise 1.2.10), so overall it is compact. According to the Bolzano-Weierstrass theorem (e.g., [17]), the sequence (x^k) therefore has at least one accumulation point in $f_{\leq}^{f(x^0)}$. In addition, the sequence $(f(x^k))$ is monotonically decreasing and bounded from below by the global minimal value of f, so it is convergent.   □

The following exercise illustrates that Lemma 2.2.3 only implies the existence of an accumulation point of the iterates x^k, but not that it is also a critical point of f.

Exercise 2.2.4 Consider the function $f(x) = x^2$, the starting point $x^0 = 3$ and the iterates $x^k = (-1)^k(1 + 1/k)$, $k \in \mathbb{N}$. Show that the iterates satisfy the descent condition from line 4 of Algorithm 2.3, that they have two accumulation points, but that none of them is a critical point of f.

To derive conditions on the choice of x^{k+1} in line 4 of Algorithm 2.3, which ensure that the algorithm terminates, we first note that without loss of generality we may assume

$$x^{k+1} = x^k + t^k d^k \qquad (2.2)$$

with some $t^k > 0$ and $d^k \in \mathbb{R}^n$. Indeed, by suitable choices of t^k and d^k, every new iterate x^{k+1} can be realized in this form. A simple possibility would be to set $d^k = x^{k+1} - x^k$ and $t^k = 1$; subsequently, we will choose d^k and t^k more cleverly.

By the approach (2.2), the choice of the new iterate x^{k+1} is divided into two separate operations, namely the determination of a *search direction d^k* and of a *step size t^k*. The classical optimization methods, which we will initially focus on, determine *first* a search direction d^k and *then* a step size t^k along the half-line defined by the search direction. Therefore, they are referred to as line search methods with step size control. The basic idea of the more modern trust-region methods, which we will discuss in Sect. 2.2.10, is to calculate *first* a search radius t^k and *then* a descent direction d^k.

We start with line search methods. In the following, let d^k be a first-order descent direction for f at x^k, i.e., $\langle \nabla f(x^k), d^k \rangle < 0$ holds. We are looking for mild conditions on the choices of t^k and d^k that guarantee $\lim_k \nabla f(x^k) = 0$, i.e., $\nabla f(x^\star) = 0$ for every accumulation point $x^\star$ of the sequence (x^k).

Due to the convergence of the sequence $(f(x^k))$ shown in Lemma 2.2.3, the differences $f(x^k + t^k d^k) - f(x^k)$ converge to zero. According to Taylor's theorem (Theorem 2.1.30), the actual descent $f(x^k + t^k d^k) - f(x^k) < 0$ roughly matches the 'first-order descent' $t^k \langle \nabla f(x^k), d^k \rangle < 0$, from which we can conclude the convergence of the vectors $\nabla f(x^k)$ to zero under suitable assumptions. First, we require that the actual descent provides a lower bound on the first-order descent, i.e., that the values of f decrease *sufficiently quickly*:

$$\exists\, c_1 > 0 \;\forall k \in \mathbb{N}: \quad f(x^k + t^k d^k) - f(x^k) \leq c_1 \cdot t^k \langle \nabla f(x^k), d^k \rangle. \qquad (2.3)$$

The sufficient decrease condition (2.3) is called *Armijo condition*. In view of $0 > f(x^k + t^k d^k) - f(x^k) \to 0$ and $c_1 \cdot t^k \langle \nabla f(x^k), d^k \rangle < 0$, the Armijo condition and the sandwich theorem yield

$$\lim_k t^k \langle \nabla f(x^k), d^k \rangle = 0. \qquad (2.4)$$

To conclude from this even $\lim_k \nabla f(x^k) = 0$, we must impose conditions on the sequences (t^k) and (d^k) that exclude the possibility that the limit in (2.4) becomes zero for other reasons.

In order to at least conclude $\lim_k \langle \nabla f(x^k), d^k \rangle = 0$, (t^k) must not converge too quickly to zero, so it must *remain sufficiently large*. We therefore require

$$\exists\, c_2 > 0 \;\forall k \in \mathbb{N}: \quad t^k \geq -c_2 \cdot \frac{\langle \nabla f(x^k), d^k \rangle}{\|d^k\|_2^2}, \qquad (2.5)$$

where we postpone the motivation for the division by $\|d^k\|_2^2$ for a moment. Because of

$$\underbrace{t^k \langle \nabla f(x^k), d^k \rangle}_{< 0,\ \to 0} \leq -c_2 \left(\frac{\langle \nabla f(x^k), d^k \rangle}{\|d^k\|_2} \right)^2 < 0$$

the sandwich theorem then yields

$$\lim_k \frac{\langle \nabla f(x^k), d^k \rangle}{\|d^k\|_2} = 0. \qquad (2.6)$$

For this result, it is even sufficient to require only the following combination of (2.3) and (2.5).

Definition 2.2.5 (Efficient Step Sizes) Let (d^k) be a sequence of first-order descent directions, and let (t^k) fulfill

$$\exists\, c > 0 \,\forall\, k \in \mathbb{N}: \quad f(x^k + t^k d^k) - f(x^k) \leq -c \cdot \left(\frac{\langle \nabla f(x^k), d^k \rangle}{\|d^k\|_2} \right)^2.$$

Then (t^k) is called an *efficient* step size sequence (for (d^k)).

The following result is proven as above with the sandwich theorem.

Theorem 2.2.6 *Let the set $f_{\leq}^{f(x^0)}$ be bounded, let (d^k) be a sequence of first-order descent directions, and let (t^k) be an efficient step size sequence. Then (2.6) applies.*

Finally we need a condition on (d^k) that guarantees $\lim_k \nabla f(x^k) = 0$. We note that (2.6) indeed holds under the desired condition $\lim_k \nabla f(x^k) = 0$ (because all vectors $d^k/\|d^k\|_2$ have length one and thus form a bounded sequence), but (2.6) can also be fulfilled when not the *lengths* of the vectors $\nabla f(x^k)$ approach zero, but in the limit their *directions* stand perpendicular to a limit point of the sequence $(d^k/\|d^k\|_2)$. In this sense, we must therefore exclude that the vectors $\nabla f(x^k)$ and $d^k/\|d^k\|_2$ stand 'asymptotically perpendicular' to each other. Without the division of d^k by $\|d^k\|_2$ this would in particular exclude the case $d^k \to 0$ which, however, would be pointless already for the simple choice $d^k = -\nabla f(x^k)$, due to the desired behavior of the sequence $(\nabla f(x^k))$.

The obtuse angles between $\nabla f(x^k)$ and $d^k/\|d^k\|_2$ should therefore not converge to a right angle. Equivalently, we may require that the negative values $\cos\big(\angle(\nabla f(x^k), d^k/\|d^k\|_2)\big)$ do not converge to zero. For this, we assume the existence of some constant $c > 0$ such that all $k \in \mathbb{N}$ satisfy the inequality

$$\cos\left(\angle \left(\nabla f(x^k), \frac{d^k}{\|d^k\|_2} \right) \right) \leq -c.$$

Due to the representation of the inner product from (2.1), this inequality is equivalent to

$$\frac{\left\langle \nabla f(x^k), \frac{d^k}{\|d^k\|_2} \right\rangle}{\left\| \frac{d^k}{\|d^k\|_2} \right\|_2} \leq -c \cdot \|\nabla f(x^k)\|_2,$$

where we have multiplied the inequality by the factor $\|\nabla f(x^k)\|_2$ to cover the case $\|\nabla f(x^k)\|_2 = 0$. This justifies the following definition.

Definition 2.2.7 (Gradient-Related Search Directions) The sequence of search directions (d^k) is called *gradient-related*, if

$$\exists\, c > 0 \,\forall\, k \in \mathbb{N}: \quad \frac{\langle \nabla f(x^k), d^k \rangle}{\|d^k\|_2} \leq -c \cdot \|\nabla f(x^k)\|_2$$

holds.

Exercise 2.2.8 Show that the sequence of search directions $d^k = -\nabla f(x^k), k \in \mathbb{N}$, is gradient-related.

We can now show the following central result for the general descent method.

Theorem 2.2.9 *Let the set $f_{\leq}^{f(x^0)}$ be bounded, and in line 4 of Algorithm 2.3 let $x^{k+1} = x^k + t^k d^k$ be chosen with a gradient-related search direction sequence (d^k) and an efficient step size sequence (t^k). For $\varepsilon = 0$, the procedure then either stops after a finite number of steps with a critical point, or the sequence (x^k) has an accumulation point, and each such point $x^\star$ satisfies $\nabla f(x^\star) = 0$.*

Proof From Lemma 2.2.3 we know that the procedure either stops after a finite number of steps with a critical point or the sequence (x^k) has an accumulation point in $f_{\leq}^{f(x^0)}$. Let $x^\star$ denote any such accumulation point. Theorem 2.2.6, the definition of gradient-relatedness, and the sandwich theorem yield the assertion.   □

That Algorithm 2.3 terminates after a finite number of steps with an exact critical point is not to be expected, but mentioned as an alternative in Theorem 2.2.9 so that one can otherwise talk about accumulation points of a 'proper' sequence.

In the case that in $f_\leq^{f(x^0)}$ there is only a single critical point $x^\star$, this must be the global minimal point of f, and every accumulation point of the sequence (x^k) from Theorem 2.2.9 coincides with $x^\star$. This means that even $\lim_k x^k = x^\star$ holds.

Theorem 2.2.9 provides the desired behavior of the general descent method for inputs with arbitrary termination tolerances $\varepsilon > 0$.

Corollary 2.2.10 *Let the set $f_\leq^{f(x^0)}$ be bounded, and in line 4 of Algorithm 2.3 let $x^{k+1} = x^k + t^k d^k$ be chosen with a gradient-related search direction sequence (d^k) and an efficient step size sequence (t^k). Then the algorithm terminates after a finite number of steps.*

2.2.2 Step Size Control

While the existence of a gradient-related search direction sequence is clear in view of Exercise 2.2.8, we still have to deal with the construction of efficient step size sequences. Whether such exist is initially unclear, since the conditions (2.3) and (2.5) roughly require upper and lower bounds on t^k at the same time and may therefore not be simultaneously accomplishable. In the following, the index $k \in \mathbb{N}$ is fixed and omitted for better clarity, i.e., we set $x = x^k$, $t = t^k$ and $d = d^k$.

In fact, we will prove the efficiency of three popular step size strategies, namely the choice of exact step sizes t_e, certain constant step sizes t_c and the Armijo step sizes t_a. In general, the choice of suitable step sizes is referred to as *step size control*.

As a basic assumption for the respective efficiency proofs, we will need the Lipschitz continuity of the gradient ∇f on the set $f_\leq^{f(x^0)}$. Recall that a function $F : D \to \mathbb{R}^m$ is called *Lipschitz continuous* on $D \subseteq \mathbb{R}^n$ (with respect to the Euclidean norm) if

$$\exists\, L > 0 \,\forall\, x, y \in D : \quad \|F(x) - F(y)\|_2 \leq L \cdot \|x - y\|_2$$

applies. Since C^1-functions are always Lipschitz continuous on compact sets (e.g., [12]), ∇f is Lipschitz continuous on the bounded set $f_\leq^{f(x^0)}$ for every C^2-function f, for example. Unfortunately, this fact does not automatically include the knowledge of the *size* of the constant L, which will cause a problem when constructing constant step sizes t_c.

Remark 2.2.11 For bounded (and therefore compact) sets $f_\leq^{f(x^0)}$, the requirement of Lipschitz continuity of ∇f on $f_\leq^{f(x^0)}$ is thus mild. This would not be the case if we had only required, for example, that f is bounded below on $\mathbb{R}^n$ (cf. Remark 2.2.1).

In the proof of the following lemma, we will need to exploit the Lipschitz continuity of ∇f on a convex set, which does not necessarily hold for the set $f_{\leq}^{f(x^0)}$. Therefore, we will even require it on the *convex hull* $\text{conv}(f_{\leq}^{f(x^0)})$ of $f_{\leq}^{f(x^0)}$, i.e., on the smallest convex superset of $f_{\leq}^{f(x^0)}$.

Remark 2.2.12 For bounded (and therefore compact) $f_{\leq}^{f(x^0)}$, the set $\text{conv}(f_{\leq}^{f(x^0)})$ is also compact, so that the requirement of Lipschitz continuity of ∇f on $\text{conv}(f_{\leq}^{f(x^0)})$ is mild.

The following result states that, for a Lipschitz continuous gradient, the qualitative error term $o(\|x - \bar{x}\|)$ from the first-order multivariate Taylor expansion (Theorem 2.1.30a) can be bounded above by an explicit quadratic term.

Lemma 2.2.13 *On a convex set $D \subseteq \mathbb{R}^n$, let f be differentiable with Lipschitz continuous gradient ∇f and associated Lipschitz constant $L > 0$. Then it holds*

$$\forall \bar{x}, x \in D : \quad |f(x) - f(\bar{x}) - \langle \nabla f(\bar{x}), x - \bar{x} \rangle| \leq \frac{L}{2} \|x - \bar{x}\|_2^2 .$$

Proof For $\bar{x}$ and x from D, due to the convexity of D, for all $t \in [0, 1]$

$$\bar{x} + t(x - \bar{x}) = (1 - t)\bar{x} + tx \in D$$

holds, so that we can exploit the Lipschitz continuity of ∇f for all points $\bar{x} + t(x - \bar{x})$ with $t \in [0, 1]$. We consider the point x as the endpoint, occurring for $t = 1$, of the line segment

$$[\bar{x}, x] = \{\bar{x} + t(x - \bar{x}) | t \in [0, 1]\}$$

and consider the error term along this line. It satisfies

$$f(\bar{x} + t(x - \bar{x})) - f(\bar{x}) - \langle \nabla f(\bar{x}), \bar{x} + t(x - \bar{x}) - \bar{x} \rangle$$
$$= f(\bar{x} + t(x - \bar{x})) - f(\bar{x}) - t\langle \nabla f(\bar{x}), x - \bar{x} \rangle.$$

Due to

$$\frac{d}{dt}(f(\bar{x} + t(x - \bar{x})) - f(\bar{x}) - t\langle \nabla f(\bar{x}), x - \bar{x} \rangle) = \langle \nabla f(\bar{x} + t(x - \bar{x})) - \nabla f(\bar{x}), x - \bar{x} \rangle$$

we can therefore write the error term at $t = 1$ 'artificially complicated' as

$$f(x) - f(\bar{x}) - \langle \nabla f(\bar{x}), x - \bar{x} \rangle = \int_0^1 \langle \nabla f(\bar{x} + t(x - \bar{x})) - \nabla f(\bar{x}), x - \bar{x} \rangle \, dt.$$

The triangle inequality for integrals, the Cauchy-Schwarz inequality, and the Lipschitz continuity of ∇f thus yield

$$
\begin{aligned}
|f(x) - f(\bar{x}) - \langle \nabla f(\bar{x}), x - \bar{x} \rangle| &\leq \int_0^1 |\langle \nabla f(\bar{x} + t(x - \bar{x})) - \nabla f(\bar{x}), x - \bar{x} \rangle|\, dt \\
&\leq \int_0^1 \|\nabla f(\bar{x} + t(x - \bar{x})) - \nabla f(\bar{x})\|_2 \cdot \|x - \bar{x}\|_2\, dt \\
&\leq \int_0^1 L \cdot t \cdot \|x - \bar{x}\|_2^2\, dt \;=\; L\|x - \bar{x}\|_2^2 \cdot \int_0^1 t\, dt \\
&= \frac{L}{2}\|x - \bar{x}\|_2^2.
\end{aligned}
$$

$\square$

Exercise 2.2.14 Show under the conditions of Lemma 2.2.13 that the condition

$$
\forall \bar{x}, x \in D: \quad f(x) - f(\bar{x}) - \langle \nabla f(\bar{x}), x - \bar{x} \rangle \;\leq\; \frac{L}{2}\|x - \bar{x}\|_2^2 \tag{2.7}
$$

is equivalent to the convexity of the function $L\|x\|_2^2/2 - f(x)$ on D. This may be shown using Theorem 2.1.40, where the assumed *continuous* differentiability of f can be weakened to differentiability [36].

The inequality (2.7) is known as the *descent lemma* (Exercise 2.2.17).

Exact Step Sizes

For $x \in f_{\leq}^{f(x^0)}$, let a first-order descent direction d for f in x be given. From $\varphi'_d(0) = \langle \nabla f(x), d \rangle < 0$ we obtain $\varphi_d(t) < \varphi_d(0)$ for all sufficiently small positive t. For bounded $f_{\leq}^{f(x^0)}$, φ_d even has global minimal points $t_e > 0$, called *exact step sizes*. By definition of the one-dimensional restriction φ_d, they satisfy

$$
f(x + t_e d) \;=\; \min_{t>0} f(x + td).
$$

An important and subsequently often used property of every exact step size, according to Fermat's rule and the chain rule, is the relationship

$$
0 \;=\; \varphi'_d(t_e) \;=\; \langle \nabla f(x + t_e d), d \rangle. \tag{2.8}
$$

Calculating an exact step size to achieve the greatest possible descent from x along d is generally intricate, so we mostly use this concept for theoretical purposes and will instead move on to *inexact* step sizes. However, for a special structure of f, exact step sizes can sometimes be easily calculated, as the following exercise shows.

Exercise 2.2.15 Consider the quadratic function $q(x) = \frac{1}{2}x^{\mathsf{T}}Ax + b^{\mathsf{T}}x$ with $A = A^{\mathsf{T}} \succ 0$ and $b \in \mathbb{R}^n$, which is coercive according to Exercise 1.2.24 and convex according to Exercise 2.1.43. Show that for every $x \in \mathbb{R}^n$ and every first-order descent direction d for q in x, the exact step size is uniquely determined by

$$t_e = -\frac{\langle Ax + b, d \rangle}{d^\mathsf{T} Ad}.$$

Theorem 2.2.16 *Let the set* $f_{\leq}^{f(x^0)}$ *be bounded, let the function* ∇f *be Lipschitz continuous on* $\mathrm{conv}(f_{\leq}^{f(x^0)})$, *and let* (d^k) *be a sequence of first-order descent directions. Then every sequence of exact step sizes* (t_e^k) *is efficient.*

Proof Let the index $k \in \mathbb{N}$ again be fixed and omitted. Since only descent steps have been performed from x^0, both x and $x + t_e d$ lie in the set $f_{\leq}^{f(x^0)}$. Let $L > 0$ be a Lipschitz constant of ∇f on $\mathrm{conv}(f_{\leq}^{f(x^0)})$. From the Cauchy-Schwarz inequality, we first obtain

$$0 = \varphi_d'(t_e) = \langle \nabla f(x + t_e d), d \rangle = \langle \nabla f(x + t_e d) - \nabla f(x), d \rangle + \langle \nabla f(x), d \rangle$$

$$\leq \|\nabla f(x + t_e d) - \nabla f(x)\|_2 \cdot \|d\|_2 + \langle \nabla f(x), d \rangle \leq L \cdot t_e \|d\|_2^2 + \langle \nabla f(x), d \rangle.$$

Therefore, t_e fulfills the condition (2.5) with $c_2 = L^{-1}$, that is,

$$t_e \geq -\frac{\langle \nabla f(x), d \rangle}{L \cdot \|d\|_2^2} =: t_c.$$

The positive auxiliary value t_c can also be considered a step size, which we will indeed do below.

Due to $x, x + t_e d \in f_{\leq}^{f(x^0)}$, the point $x + td$ lies in $\mathrm{conv}(f_{\leq}^{f(x^0)})$ for all $t \in (0, t_e]$. Lemma 2.2.13 therefore provides for these t

$$f(x + td) - f(x) - t\langle \nabla f(x), d \rangle \leq |f(x + td) - f(x) - t\langle \nabla f(x), d \rangle|$$

$$= t^2 \frac{L}{2} \|d\|_2^2.$$

For t_c it follows

$$f(x + t_c d) - f(x) \leq t_c \left(\langle \nabla f(x), d \rangle + t_c \frac{L}{2} \|d\|_2^2 \right)$$

$$= -\frac{\langle \nabla f(x), d \rangle}{L \cdot \|d\|_2^2} \left(\langle \nabla f(x), d \rangle - \frac{\langle \nabla f(x), d \rangle}{L \cdot \|d\|_2^2} \cdot \frac{L}{2} \|d\|_2^2 \right)$$

$$= -\frac{1}{2L} \left(\frac{\langle \nabla f(x), d \rangle}{\|d\|_2} \right)^2.$$

Because of $t_c > 0$ and the global minimality of the exact step size t_e for $\varphi_d(t)$ over $t > 0$ we finally obtain

$$f(x + t_e d) - f(x) \leq f(x + t_c d) - f(x) \leq -\frac{1}{2L}\left(\frac{\langle \nabla f(x), d \rangle}{\|d\|_2}\right)^2,$$

so that with the iteration-independent constant $c = (2L)^{-1}$ the claim follows. $\square$

In addition the existence of gradient-related search directions this shows also the existence of efficient step sizes.

Constant Step Sizes

If the function f does not have a special structure (as in Exercise 2.2.15), it is usually not worth the effort to calculate an exact step size t_e^k in each iteration step. Since one is less interested in minimal points of the auxiliary functions φ_{d^k} than in those of f, one then prefers to use *inexact* step sizes, which are also efficient, but easier to calculate.

A seemingly obvious possibility for this is to use, instead of t_e^k, the auxiliary quantities that occurred in the proof of Theorem 2.2.16,

$$t_c^k = -\frac{\langle \nabla f(x^k), d^k \rangle}{L \cdot \|d^k\|_2^2}.$$

They can be interpreted as step sizes, and the efficiency of the corresponding sequence (t_c^k) was a main ingredient of the above proof. In the special case $d^k = -\nabla f(x^k)$ we even have

$$t_c^k = \frac{1}{L},$$

so that the sequence of step sizes is then *constant*.

Unfortunately, this choice can only be implemented algorithmically if a Lipschitz constant $L > 0$ of ∇f on $\mathrm{conv}(f_{\leq}^{f(x^0)})$ is *explicitly known*. With some effort, e.g. using interval arithmetic, Lipschitz constants can actually often be determined [36]. Then, however, while only sufficiently small Lipschitz constants describe the behavior of ∇f well, one typically has to be content with rough overestimates of L. This results in small corresponding step sizes t_c^k, so that the iteration progresses slower than necessary.

We emphasize that the explicit knowledge of some L is not required in the efficiency proof for the exact step sizes in Theorem 2.2.16.

Exercise 2.2.17 Let the set $f_{\leq}^{f(x^0)}$ be bounded, let the function ∇f be Lipschitz continuous on $\mathrm{conv}(f_{\leq}^{f(x^0)})$ with Lipschitz constant $L > 0$, and in Algorithm 2.3 let the search direction $d^k = -\nabla f(x^k)$ of the gradient method be chosen at the iterate x^k. Using inequality (2.7), show that every step size $t^k \in (0, 2/L)$ leads to the descent

$$f(x^{k+1}) \leq f(x^k) - t^k(1 - \tfrac{t^k}{2}L)\|\nabla f(x^k)\|_2^2 < f(x^k),$$

where the factor $t^k(1 - \tfrac{t^k}{2}L)$ is maximal for $t_c^k = 1/L$. This relationship motivates to call (2.7) descent lemma.

Armijo Step Sizes

A popular inexact step size control in modern implementations of optimization algorithms goes back to an idea by Armijo: For $x \in f_{\leq}^{f(x^0)}$, let d be a first-order descent direction and $\sigma \in (0, 1)$. Then there exists some $\check{t} > 0$, such that for all $t \in (0, \check{t})$ the values $\varphi_d(t)$ lie below the 'upwards rotated tangent' $\varphi_d(0) + t\sigma\varphi_d'(0)$, so that

$$f(x + td) \leq f(x) + t\sigma\langle\nabla f(x), d\rangle$$

holds (Fig. 2.5).

Every such t satisfies the Armijo condition (2.3) with $c_1 = \sigma$. But how can one choose some $t_a > 0$ among these step sizes that also satisfies (2.5)? This is realized by the *Armijo rule* given in Algorithm 2.4 with a *backtracking line search* idea.

> **Theorem 2.2.18** *Let the set* $f_{\leq}^{f(x^0)}$ *be bounded, let the function* ∇f *be Lipschitz continuous on* $\mathrm{conv}(f_{\leq}^{f(x^0)})$, *and let* (d^k) *be a sequence of first-order descent directions. Then the sequence of Armijo step sizes* (t_a^k) *from Algorithm 2.4 (with parameters* σ, ρ *and* γ *chosen independently of k) is well-defined and efficient.*

Proof Let the index $k \in \mathbb{N}$ be again fixed and omitted. Because of $\rho \in (0, 1)$, in Algorithm 2.4 we have $\lim_\ell t^\ell = 0$, and there exists some $\ell_0 \in \mathbb{N}$ with $t^{\ell_0} \in (0, \check{t})$ (Fig. 2.5). The termination

Fig. 2.5 Armijo rule

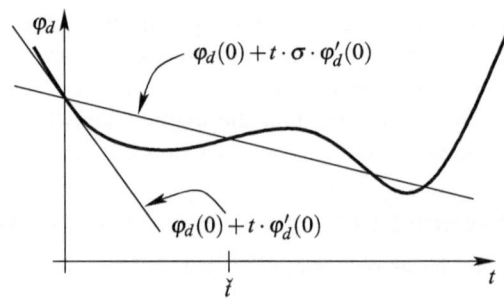

Algorithm 2.4: Armijo rule

Input: C^1-function f, parameters $\sigma, \rho \in (0, 1)$, $\gamma > 0$ as well as $x, d \in \mathbb{R}^n$ with
$\langle \nabla f(x), d \rangle < 0$

Output: Armijo step size t_a

1 **begin**
2   Choose a starting step size $t^0 \geq -\gamma \langle \nabla f(x), d \rangle / \|d\|_2^2$ and set $\ell = 0$.
3   **while** $f(x + t^\ell d) > f(x) + t^\ell \sigma \langle \nabla f(x), d \rangle$ **do**
4   Set $t^{\ell+1} = \rho t^\ell$.
5   Replace ℓ with $\ell + 1$.
6   **end**
7   Set $t_a = t^\ell$.
8 **end**

---

condition in line 3 of Algorithm 2.4 is therefore met after a finite number of steps, and the sequence of Armijo step sizes is thus well-defined.

Since, by the above preliminary considerations, (2.3) with $c_1 = \sigma$ is fulfilled for t_a, (2.5) remains to be shown. If Algorithm 2.4 already terminates at $\ell = 0$, then (2.5) holds with $c_2 = \gamma$. In the following, let $t_a = t^\ell$ with $\ell > 0$. Since $t^{\ell-1}$ did not yet meet the termination criterion, it holds

$$f(x + t^{\ell-1} d) > f(x) + t^{\ell-1} \sigma \langle \nabla f(x), d \rangle. \tag{2.9}$$

To apply the assumed Lipschitz condition on $\mathrm{conv}(f_{\leq}^{f(x^0)})$, we distinguish two cases, where $[x, x + t^{\ell-1} d]$ denotes the line segment $\{ x + t d \mid t \in [0, t^{\ell-1}] \}$.

Case 1: $[x, x + t^{\ell-1} d] \subseteq \mathrm{conv}(f_{\leq}^{f(x^0)})$. According to the mean value theorem, there exists some $\theta \in (0, 1)$ with

$$f(x + t^{\ell-1} d) = f(x) + t^{\ell-1} \langle \nabla f(x + \theta t^{\ell-1} d), d \rangle.$$

Due to (2.9) and $t^{\ell-1} > 0$, it follows

$$\sigma \langle \nabla f(x), d \rangle < \langle \nabla f(x + \theta t^{\ell-1} d), d \rangle,$$

thus by means of the Cauchy-Schwarz inequality and the Lipschitz condition

$$(\sigma - 1) \langle \nabla f(x), d \rangle < \langle \nabla f(x + \theta t^{\ell-1} d) - \nabla f(x), d \rangle \leq L \cdot \theta t^{\ell-1} \|d\|_2^2 \leq L t^{\ell-1} \|d\|_2^2.$$

We obtain

$$t_a = \rho \cdot t^{\ell-1} \geq -\frac{\rho(1 - \sigma)}{L} \frac{\langle \nabla f(x), d \rangle}{\|d\|_2^2}.$$

Case 2: $[x, x + t^{\ell-1} d] \not\subseteq \mathrm{conv}(f_{\leq}^{f(x^0)})$. The Lipschitz estimate is not guaranteed on the entire segment $[x, x + t^{\ell-1} d]$ in this case. However, with an *exact* step size t_e, both x and $x + t_e d$ lie in $f_{\leq}^{f(x^0)}$ and therefore the line segment $[x, x + t_e d]$ is a subset of $\mathrm{conv}(f_{\leq}^{f(x^0)})$. The condition of the second case therefore implies $t_e \leq t^{\ell-1}$. With the step size $t_c \leq t_e$ from the proof of

Theorem 2.2.16, it follows

$$t_a = \rho \, t^{\ell-1} \geq \rho \, t_e \geq \rho \, t_c = -\frac{\rho}{L} \frac{\langle \nabla f(x), d \rangle}{\|d\|_2^2} \geq -\frac{\rho(1-\sigma)}{L} \frac{\langle \nabla f(x), d \rangle}{\|d\|_2^2}.$$

Overall, (2.5) is therefore satisfied in both cases with

$$c_2 = \min \left\{ \gamma, \ \frac{\rho(1-\sigma)}{L} \right\}.$$

Since c_1 and c_2 are iteration-independent, the claim follows.   □

Note that the specific size of the Lipschitz constant L is also irrelevant for this efficiency proof.

In practice, values $\sigma \in [0.01, 0.2]$ and $\rho = 0.5$ have proven to be effective. Instead of determining t^0 in Algorithm 2.4 by γ, for simplicity it is often set to $t^0 := 1$. The following exercise shows that Algorithm 2.3 does not necessarily terminate in this case.

Exercise 2.2.19 Show for the function $f(x) = \frac{1}{2}x^2$, the starting point $x^0 = -3$, the directions $d^k = 2^{-k}$ as well as $\sigma = \frac{1}{2}$, that the modification of Algorithm 2.4 by the choice $t^0 := 1$ does not lead to an efficient sequence of step sizes.

One should therefore initialize t^0 as indicated in Algorithm 2.4 where, e.g., the choice $\gamma = 10^{-4}$ has proven to be effective. It is also not difficult to see that the Armijo rule may as well be used for only one-sided directionally differentiable functions, by replacing the inner product $\langle \nabla f(\bar{x}), d \rangle$ with $f'(\bar{x}, d)$ in lines 2 and 3 of the algorithm. We will employ this generalization in Sect. 3.3.6.

2.2.3 Gradient Method

After these preparations, we can state a first implementable minimization method, namely the *gradient method* (Algorithm 2.5). It is also known under the names *Cauchy method* and, due to its basic geometric idea (Sect. 2.1.3), *steepest descent method.*

Theorem 2.2.20 *Let the set $f_{\leq}^{f(x^0)}$ be bounded, let the function ∇f be Lipschitz continuous on $\mathrm{conv}(f_{\leq}^{f(x^0)})$, and let line 5 employ exact step sizes (t_e^k) or Armijo step sizes (t_a^k). Then Algorithm 2.5 terminates after finitely many steps. If a Lipschitz constant $L > 0$ for the Lipschitz continuity of ∇f on $\mathrm{conv}(f_{\leq}^{f(x^0)})$ is known, then this result also applies to the then computable constant step sizes $t_c^k = L^{-1}, k \in \mathbb{N}$.*

Algorithm 2.5: Gradient method

Input: C^1-optimization problem P, starting point x^0 and termination tolerance $\varepsilon > 0$

Output: Approximation $\bar{x}$ of a critical point of f (if the method terminates; Theorem 2.2.20)

```
1 begin
2     Set k = 0.
3     while ‖∇f(x^k)‖ > ε do
4         Set d^k = −∇f(x^k).
5         Determine a step size t^k.
6         Set x^{k+1} = x^k + t^k d^k.
7         Replace k with k + 1.
8     end
9     Set x̄ = x^k.
10 end
```

Proof The assertions follow from Corollary 2.2.10, Exercise 2.2.8 and Theorems 2.2.16 and 2.2.18. □

When applied to convex-quadratic objective functions f, even better convergence results for the gradient method can be derived. In preparation for this, we recall the definition of the *spectral norm* of a (not necessarily square) matrix A,

$$\|A\|_2 := \max\{\|Ad\|_2 \mid \|d\|_2 = 1\}.$$

From this definition, for every vector $d \neq 0$, the estimate

$$\|Ad\|_2 = \left\| A \frac{d}{\|d\|_2} \right\|_2 \|d\|_2 \leq \|A\|_2 \|d\|_2 \qquad (2.10)$$

follows (for $d = 0$ it holds anyway).

Exercise 2.2.21 Consider the quadratic function $q(x) = \frac{1}{2}x^\mathsf{T} Ax + b^\mathsf{T} x$ with $A = A^\mathsf{T}$ and $b \in \mathbb{R}^n$. Show that the gradient ∇q is Lipschitz continuous on $\mathbb{R}^n$ with $L = \|A\|_2$.

Example 2.2.22 For the function $q(x) = \frac{1}{2}x^\mathsf{T} Ax + b^\mathsf{T} x$ with $A = A^\mathsf{T} \succ 0$ and $b \in \mathbb{R}^n$, the gradient method generates a sequence of iterates (x^k) that even converges to the global minimal point of q when either exact, constant or Armijo step sizes are chosen.

Indeed, q is coercive according to Exercise 1.2.24, so for every $x^0 \in \mathbb{R}^n$ the lower level set $q_{\leq}^{q(x^0)}$ is bounded. Furthermore, ∇q is Lipschitz continuous on all of $\mathbb{R}^n$ with $L = \|A\|_2$ according to Exercise 2.2.21. The mentioned step size controls are therefore efficient, so that according to Theorem 2.2.9 every accumulation point of the iterates x^k generated by the gradient method for q is a critical point of q.

Exercise 2.1.43 also shows that the *only* critical point $x^\star = -A^{-1}b$ of q coincides with the unique global minimal point. Thus, the sequence of iterates (x^k) converges, and its limit point is the global minimal point of q.

According to Exercise 2.2.15, for every first-order descent direction for q at x, the unique exact step size is

$$t_e = -\frac{\langle Ax + b, d \rangle}{d^\intercal A d}.$$

For the gradient method this yields

$$t_e = \frac{\|\nabla q(x)\|_2^2}{Dq(x)A\nabla q(x)}$$

with $\nabla q(x) = Ax + b$.

Since the Lipschitz constant $L = \|A\|_2$ is explicitly known, the choice of the constant step size $t_c = \|A\|_2^{-1}$ is also possible. We will deal with the calculation of the spectral norm $\|A\|_2$ of a symmetric and positive definite matrix A as the largest eigenvalue $\lambda_{\max}(A)$ of A in Remark 2.2.41.

In view of the convergence results, and since it is easy to implement, the gradient method is popular in practice. However, the strategy of choosing the local steepest descent in each iteration (a *greedy strategy*) has the disadvantage that the method often converges *very slowly*. Then only rough tolerances $\varepsilon > 0$ can be specified if the method is to terminate within a reasonable time limit.

Geometrically, the slowness of the gradient method can be explained with the help of the results from Sect. 2.1.3 (Fig. 2.6): If the level lines of f have the shape of elongated ellipses with a minimal point $x^\star$ in their common center, then $-\nabla f(x^k)$ typically does not point in the direction of $x^\star$. The iterates therefore move along a 'zigzag line', which is referred to as the *zigzagging effect*.

To construct better methods, we first need a classification of convergence speeds.

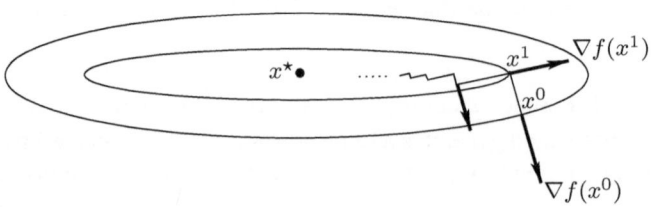

Fig. 2.6 Zigzagging effect

Definition 2.2.23 (Speeds of Convergence) Let (x^k) be a convergent sequence with limit point $x^\star$. It is called

(a) *linearly convergent*, if

$$\exists\, 0 < c < 1,\ k_0 \in \mathbb{N}\quad \forall\, k \geq k_0:\quad \|x^{k+1} - x^\star\| \leq c \cdot \|x^k - x^\star\|,$$

(b) *superlinearly convergent*, if

$$\exists\, c^k \searrow 0,\ k_0 \in \mathbb{N}\quad \forall\, k \geq k_0:\quad \|x^{k+1} - x^\star\| \leq c^k \cdot \|x^k - x^\star\|,$$

(c) *quadratically convergent*, if

$$\exists\, c > 0,\ k_0 \in \mathbb{N}\quad \forall k \geq k_0:\quad \|x^{k+1} - x^\star\| \leq c \cdot \|x^k - x^\star\|^2.$$

It is not difficult to see that quadratic convergence implies superlinear convergence and that the latter implies linear convergence. While superlinear convergence is 'fast' and quadratic convergence is 'very fast', linear convergence can be very slow, in particular for a constant $c \approx 1$.

We emphasize that the points x^k in Definition 2.2.23 do not necessarily have to be iterates of some descent method. In particular, one can also measure the convergence speed of the function values $f(x^k)$ towards a limit $f(x^\star)$ for a descent method.

Indeed, the following theorem shows that, even for 'very pleasant' functions, the gradient method generates iterates whose objective function values converge only linearly, and with a constant c that can be close to one. To this end, we consider the convex-quadratic function $q(x) = \frac{1}{2}x^\mathsf{T} A x + b^\mathsf{T} x$ with $A = A^\mathsf{T} \succ 0$ and $b \in \mathbb{R}^n$ and denote the largest and smallest eigenvalue of the matrix A by $\lambda_{\max}$ and $\lambda_{\min}$, respectively. According to Example 2.2.22, the iterates of the gradient method with exact step sizes converge towards the global minimal point $x^\star = -A^{-1}b$ of q, and the continuity of q implies the convergence of the function values $q(x^k)$ towards $q(x^\star) = -\frac{1}{2}b^\mathsf{T} A^{-1}b$. Next, we consider the convergence *speed* of these function values.

For this, we need the following result, which is proven in [16], for example.

Lemma 2.2.24 (Kantorovich Inequality) *Let $A = A^\mathsf{T} \succ 0$ with maximal and minimal eigenvalue $\lambda_{\max}$ and $\lambda_{\min}$, respectively. Then for every $v \in \mathbb{R}^n \setminus \{0\}$ the estimate*

$$\frac{v^\mathsf{T} A^{-1} v \cdot v^\mathsf{T} A v}{\|v\|_2^4} \ \leq\ \frac{(\lambda_{\max} + \lambda_{\min})^2}{4\,\lambda_{\max}\,\lambda_{\min}}$$

holds.

Theorem 2.2.25 *Let the gradient method with exact step sizes and $\varepsilon = 0$ be applied to the convex-quadratic function $q(x) = \frac{1}{2}x^{\mathsf{T}}Ax + b^{\mathsf{T}}x$ with $A = A^{\mathsf{T}} \succ 0$ and $b \in \mathbb{R}^n$. Then for all $k \in \mathbb{N}$ the estimate*

$$|q(x^{k+1}) - q(x^{\star})| \leq \left(\frac{\lambda_{\max} - \lambda_{\min}}{\lambda_{\max} + \lambda_{\min}}\right)^2 |q(x^k) - q(x^{\star})|$$

holds.

Proof Let $k \in \mathbb{N}$ be fixed. For clarity, we omit the index k in the following and set $x^+ := x^{k+1}$. Since $q(x^{\star})$ is the global minimal value of q, the absolute values in the claimed inequality are superfluous. Due to $x^+ = x - t_e \nabla q(x)$ and

$$t_e = \frac{\|\nabla q(x)\|_2^2}{Dq(x)A\nabla q(x)}$$

it holds

$$q(x^+) - q(x^{\star}) = q(x) - t_e\|\nabla q(x)\|_2^2 + \frac{t_e^2}{2} Dq(x)A\nabla q(x) - q(x^{\star})$$

$$= q(x) - q(x^{\star}) - \frac{1}{2}\frac{\|\nabla q(x)\|_2^4}{Dq(x)A\nabla q(x)}$$

and

$$q(x) - q(x^{\star}) = \frac{1}{2} Dq(x)A^{-1}\nabla q(x),$$

as can be verified by expanding the term on the right side. It follows

$$\frac{q(x^+) - q(x^{\star})}{q(x) - q(x^{\star})} = 1 - \frac{\frac{1}{2}\frac{\|\nabla q(x)\|_2^4}{Dq(x)A\nabla q(x)}}{q(x) - q(x^{\star})} = 1 - \frac{\|\nabla q(x)\|_2^4}{Dq(x)A^{-1}\nabla q(x) \cdot Dq(x)A\nabla q(x)}.$$

Since the gradient method would have terminated before the calculation of x^+ in the case $\nabla q(x) = 0$, Lemma 2.2.24 can be applied with $v = \nabla q(x) \neq 0$, and we obtain

$$\frac{q(x^+) - q(x^{\star})}{q(x) - q(x^{\star})} \leq 1 - \frac{4\lambda_{\max}\lambda_{\min}}{(\lambda_{\max} + \lambda_{\min})^2} = \left(\frac{\lambda_{\max} - \lambda_{\min}}{\lambda_{\max} + \lambda_{\min}}\right)^2.$$

$\square$

Theorem 2.2.25 only states that the objective function values of the iterates in the gradient method converge *at least* linearly with the constant

$$c = \left(\frac{\lambda_{\max} - \lambda_{\min}}{\lambda_{\max} + \lambda_{\min}}\right)^2$$

under the given conditions, which does not exclude that the method may actually run faster. However, numerical experience with the gradient method shows that indeed only linear convergence with the calculated constant can be expected. The crucial point here is that, by choosing the matrix A appropriately, constants c arbitrarily close to one can be generated, thus achieving arbitrarily slow linear convergence of the gradient method. Since c is close to one if $\lambda_{\min}$ is small compared to $\lambda_{\max}$, the geometric reason for slow convergence are 'very elongated' level sets of q, as already illustrated as level lines in Fig. 2.6.

Remark 2.2.26 (Ellipsoidal Level Sets and Eigenvalues) Since understanding the relationship between eigenvalues and level sets is also essential for the following section, we briefly recall some basics about the ellipsoidal level sets

$$q_{=}^{\alpha} := \{x \in \mathbb{R}^n \mid q(x) = \alpha\}$$

of convex-quadratic functions $q(x) = \frac{1}{2}x^{\mathsf{T}}Ax + b^{\mathsf{T}}x$ with $A = A^{\mathsf{T}} \succ 0$ and $b \in \mathbb{R}^n$.

According to Exercise 2.1.43, $\alpha_{\min} = -\frac{1}{2}b^{\mathsf{T}}A^{-1}b$ is the minimal level of q, and the corresponding minimal point $x^{\star} = -A^{-1}b$ can be considered as the center of each ellipsoid $q_{=}^{\alpha}$ with $\alpha > \alpha_{\min}$. For each such α, one can move from this center $x^{\star}$ in any given direction $d \in \mathbb{R}^n$ and will certainly hit the level set $q_{=}^{\alpha}$. The size of the corresponding step $t > 0$ with $x^{\star} + td \in q_{=}^{\alpha}$ generally depends on the direction d. Expressed in formulas, this is the question for the $t > 0$ which, given d, satisfies

$$
\begin{aligned}
\alpha = q(x^{\star} + td) &= \tfrac{1}{2}(x^{\star} + td)^{\mathsf{T}}A(x^{\star} + td) + b^{\mathsf{T}}(x^{\star} + td) \\
&= \tfrac{1}{2}(x^{\star})^{\mathsf{T}}A(x^{\star}) + td^{\mathsf{T}}Ax^{\star} + \tfrac{1}{2}t^2 d^{\mathsf{T}}Ad + b^{\mathsf{T}}x^{\star} + tb^{\mathsf{T}}d \\
&= \left(\tfrac{1}{2}(x^{\star})^{\mathsf{T}}A(x^{\star}) + b^{\mathsf{T}}x^{\star}\right) + td^{\mathsf{T}}A(-A^{-1}b) + \tfrac{1}{2}t^2 d^{\mathsf{T}}Ad + tb^{\mathsf{T}}d \\
&= -\tfrac{1}{2}b^{\mathsf{T}}A^{-1}b - td^{\mathsf{T}}b + \tfrac{1}{2}t^2 d^{\mathsf{T}}Ad + tb^{\mathsf{T}}d \\
&= \alpha_{\min} + \tfrac{1}{2}t^2 d^{\mathsf{T}}Ad.
\end{aligned}
$$

If we choose as direction d specifically a unit length eigenvector v of the matrix A, then with the associated eigenvalue λ we further obtain

$$\alpha - \alpha_{\min} = \tfrac{1}{2}t^2 v^{\mathsf{T}}Av = \tfrac{1}{2}t^2 v^{\mathsf{T}}(\lambda v) = \tfrac{1}{2}\lambda t^2 \|v\|_2^2 = \tfrac{1}{2}\lambda t^2$$

and thus

$$t = \sqrt{\frac{2(\alpha - \alpha_{\min})}{\lambda}},$$

where we have used that λ is positive due to the positive definiteness of A. The derived formula for t states that in the direction of an eigenvector v of A (a *principal axis* of the ellipsoid), it depends on the square root of the reciprocal of the associated eigenvalue when the set $q_{\leq}^{\alpha}$ is hit. For eigenvalues close to zero, the path is thus very long, while it is short for large eigenvalues. The length of such a path is also referred to as the length of the *semi-axis* of the associated ellipsoid determined by v.

This finally explains that a large discrepancy between $\lambda_{\min}$ and $\lambda_{\max}$ corresponds to a large discrepancy between the longest and shortest semi-axis of each ellipsoid that forms a level set of q, i.e., that the level sets of q are geometrically 'very elongated'.

Exercise 2.2.27 For a matrix $A = A^{\mathsf{T}} \succ 0$, calculate the lengths of the semi-axes of the ellipsoid $\{x \in \mathbb{R}^n \mid x^{\mathsf{T}} A x = 1\}$.

Exercise 2.2.28 For a matrix $A = A^{\mathsf{T}} \succ 0$ with maximal and minimal eigenvalues $\lambda_{\max}$ and $\lambda_{\min}$, respectively, $\kappa := \lambda_{\max}/\lambda_{\min}$ is called the *condition number* of A. If the condition number is close to one, one speaks of a *well-conditioned* matrix. Express the linearity factor of the gradient method from Theorem 2.2.25 using the condition number. How does the condition of A affect the convergence speed of the method?

2.2.4   Variable Metric Methods

Theorem 2.2.25 on the slow convergence of the gradient method and its geometric interpretation (Fig. 2.6) suggest the idea of replacing the descent direction $d^k = -\nabla f(x^k)$ with a direction that takes *curvature* information about f into account. This can be accomplished as follows.

According to Theorem 2.2.25, the gradient method (with exact step sizes) minimizes a convex-quadratic function $q(x) = \frac{1}{2} x^{\mathsf{T}} A x + b^{\mathsf{T}} x$ in a *single* step if the smallest and largest eigenvalue $\lambda_{\min}$ and $\lambda_{\max}$ of A coincide. Then also *all* eigenvalues of A coincide, so that q has spherical level sets.

The main geometric idea of the following methods is to introduce a *new coordinate system* at each iterate x^k when minimizing a (not necessarily convex-quadratic) C^1-function f, such that f has level sets which 'look as spherical as possible' around x^k in the new coordinates. In these new coordinates, a descent in the negative gradient direction is therefore sensible. If the coordinate transformation is *linear*, we will be able to derive a simple representation of this search direction in the original coordinates, so that the explicit use of the coordinate transformation is then no longer necessary.

Figure 2.7 illustrates the construction using the example of a convex-quadratic function $q(x) = \frac{1}{2} x^{\mathsf{T}} A x + b^{\mathsf{T}} x$, where we neglect that the point at which the search direction is to be calculated is an iterate, and instead denote it with $\bar{x}$. On the left, elliptical level lines of q are shown, for which the gradient method would lead to slow convergence from $\bar{x}$. A coordinate system that 'fits better' to the level lines in terms of both orientation and scaling is shown in dashed lines, and the new coordinates are denoted by y_1 and y_2 instead of x_1 and x_2. One could additionally

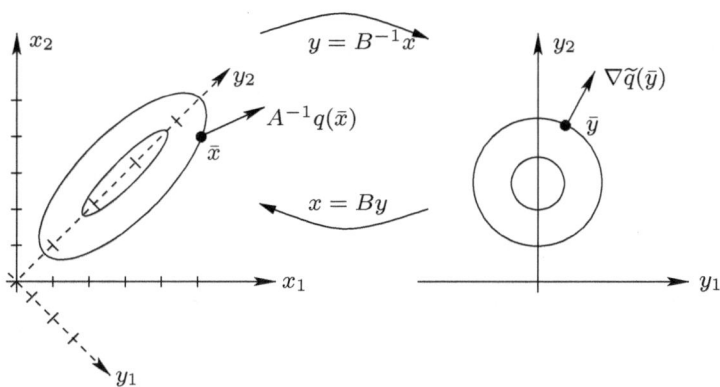

Fig. 2.7 Coordinate transformation in the variable metric idea

choose the common center of the ellipses as the origin of the new coordinate system, but this would turn out to be unnecessary.

For this specific coordinate system, one can read from the figure the possible new 'unit vectors'

$$b^1 = \begin{pmatrix} \frac{1}{2} \\ -\frac{1}{2} \end{pmatrix} \quad \text{and} \quad b^2 = \begin{pmatrix} 1 \\ 1 \end{pmatrix}.$$

In the language of linear algebra, b^1 and b^2 are *basis vectors* of the new coordinate system. With their help, one can determine the coordinates y_1 and y_2 of any point $x \in \mathbb{R}^2$ with respect to the new coordinate system, as these satisfy the linear system of equations

$$x = y_1 b^1 + y_2 b^2 = By$$

with the matrix $B := (b^1, b^2)$. For example, the new coordinates y_1 and y_2 of the point $x = (1, 0)^\mathsf{T}$ satisfy the system

$$\begin{pmatrix} 1 \\ 0 \end{pmatrix} = \begin{pmatrix} \frac{1}{2} & 1 \\ -\frac{1}{2} & 1 \end{pmatrix} \begin{pmatrix} y_1 \\ y_2 \end{pmatrix},$$

from which $y_1 = 1$ and $y_2 = \frac{1}{2}$ follows. Indeed, to determine new coordinates, it is not necessary to solve systems of equations. Since the vectors b^1 and b^2 form a basis of $\mathbb{R}^2$, the matrix B is invertible, and the formula $y = B^{-1}x$ explicitly provides the new coordinates y for every $x \in \mathbb{R}^2$.

The mapping $x \mapsto B^{-1}x$ is the sought-after linear coordinate transformation. It transforms the geometry of the left side from Fig. 2.7 into that of the right side.

In particular, the point $\bar{x}$, at which we are looking for a descent direction superior to the negative gradient of q, is mapped to the point $\bar{y}$ by the transformation. In the new coordinates y, the level lines of q are circular, so that from $\bar{y}$ the negative gradient direction points to the minimal point. The crucial question is how the 'function q in new coordinates' is explicitly defined, since of this function the level lines are considered, and its gradient is to be calculated. Because of

$$q(x) \;=\; q(By) \;=\; (q \circ B)(y)$$

the function $\widetilde{q} := q \circ B$ is the sought representation of q in new coordinates, and $\nabla \widetilde{q}(\bar{y})$ is the sought gradient direction in $\bar{y}$.

In a final step, we want to transform the direction found in the new coordinates back into the original coordinates, where the reverse transformation of the geometry of the right side from Fig. 2.7 into that of the left side is realized by the mapping $y \mapsto By$. To represent the resulting vector $B\nabla \widetilde{q}(\bar{y})$ in original coordinates, we use the chain rule and obtain for each $y \in \mathbb{R}^2$

$$D\widetilde{q}(y) \;=\; D[q(B(y))] \;=\; Dq(By)B \;=\; Dq(x)B,$$

so in particular $\nabla \widetilde{q}(\bar{y}) = B^\mathsf{T}\nabla q(\bar{x})$ and

$$B\nabla \widetilde{q}(\bar{y}) \;=\; BB^\mathsf{T}\nabla q(\bar{x}) \;=\; \frac{1}{4}\begin{pmatrix} 5 & 3 \\ 3 & 5 \end{pmatrix}\nabla q(\bar{x}).$$

For the general case $n \geq 1$ and without graphical illustration, at this point it remains unclear how the matrix B should be constructed. Fortunately, also in general there always exists a right-angled coordinate system that is 'appropriately aligned' to the orientation of the ellipsoidal level sets of q. Its axes correspond exactly to the principal axes of these ellipsoids, i.e., the directions given by the eigenvectors $v^1, \ldots, v^n$ of A (since A is symmetric, there are indeed exactly n different and pairwise orthogonal eigenvectors of A). As in Fig. 2.7, however, we refrain from identifying the center of the ellipsoids (i.e., the minimal point of q) with the origin of the new coordinate system.

We also need a 'suitable scaling' of the axes to obtain spherical level sets in the new coordinate system. For this purpose, we choose eigenvectors $v^1, \ldots, v^n$ of A normalized to length one and look for suitable (positive) scaling factors for the individual vectors. More explicitly, the new coordinate system should possess the vectors $b^i := c_i v^i$, $i = 1, \ldots, n$, as 'unit vectors', with suitably chosen factors $c_1, \ldots, c_n > 0$. The new coordinates $y_1, \ldots, y_n$ of a point $x \in \mathbb{R}^n$ then satisfy the system

$$x \;=\; y_1 b^1 + \ldots + y_n b^n \;=\; y_1 c_1 v^1 + \ldots + y_n c_n v^n \;=\; V\begin{pmatrix} c_1 y_1 \\ \vdots \\ c_n y_n \end{pmatrix},$$

where we have introduced the matrix $V := (v^1, \ldots, v^n)$. With the diagonal matrix

$$C := \begin{pmatrix} c_1 & & \\ & \ddots & \\ & & c_n \end{pmatrix}$$

the system can be written even more briefly as $x = VCy$, and for the matrix $B := (b^1, \ldots, b^n)$ the representation $B = VC$ can be read off. In new coordinates, the function q is thus $\tilde{q}(y) = q(By)$ with $B = VC$.

To calculate the scaling factors $c_1, \ldots, c_n$ which lead to spherical level sets of $\tilde{q}$, we ensure that the Hessian matrix of $\tilde{q}$ has identical eigenvalues. In view of

$$\tilde{q}(y) = q(By) = \tfrac{1}{2}(By)^\mathsf{T} A(By) + b^\mathsf{T}(By) = \tfrac{1}{2}y^\mathsf{T} B^\mathsf{T} ABy + b^\mathsf{T} By$$

the Hessian is

$$D^2\tilde{q}(y) = B^\mathsf{T} AB = CV^\mathsf{T} AVC.$$

In the next step, the *spectral theorem* helps us, according to which the matrix $V^\mathsf{T} AV$ appearing in $D^2\tilde{q}(y)$ coincides with the diagonal matrix

$$\Lambda = \begin{pmatrix} \lambda_1 & & \\ & \ddots & \\ & & \lambda_n \end{pmatrix}$$

of the eigenvalues of A. This is not hard to see, because the normalization and mutual orthogonality of the eigenvectors imply $(v^i)^\mathsf{T} v^i = \|v^i\|_2^2 = 1$ for all $i = 1, \ldots, n$ and $(v^i)^\mathsf{T} v^j = 0$ for all $i \neq j$. Therefore,

$$V^\mathsf{T} AV = V^\mathsf{T}(Av^1, \ldots, Av^n) = \begin{pmatrix} (v^1)^\mathsf{T} \\ \vdots \\ (v^n)^\mathsf{T} \end{pmatrix} (\lambda_1 v^1, \ldots, \lambda_n v^n) = \Lambda$$

and thus

$$D^2\tilde{q}(y) = CV^\mathsf{T} AVC = C\Lambda C = \begin{pmatrix} \lambda_1 c_1^2 & & \\ & \ddots & \\ & & \lambda_n c_n^2 \end{pmatrix}$$

hold. So $\widetilde{q}$ has spherical level sets for, e.g., the choices $c_i := 1/\sqrt{\lambda_i}$, $i = 1, \ldots, n$. With the definitions

$$\Lambda^{\frac{1}{2}} := \begin{pmatrix} \sqrt{\lambda_1} & & \\ & \ddots & \\ & & \sqrt{\lambda_n} \end{pmatrix}$$

and $\Lambda^{-\frac{1}{2}} := (\Lambda^{\frac{1}{2}})^{-1}$ this relationship can also be written briefly as $C = \Lambda^{-\frac{1}{2}}$, so that the sought matrix is

$$B = VC = V\Lambda^{-\frac{1}{2}}.$$

As the transformation of the gradient vector $\nabla \widetilde{q}(\bar{y})$ back to original coordinates we obtain as above the vector $BB^{\mathsf{T}} \nabla q(\bar{x})$, where the matrix BB^{T} can now be written as

$$BB^{\mathsf{T}} = VC(VC)^{\mathsf{T}} = VC^2 V^{\mathsf{T}} = V\Lambda^{-1} V^{\mathsf{T}}.$$

To see that this matrix is identical to A^{-1}, we invert both sides of the spectral theorem equation $V^{\mathsf{T}} A V = \Lambda$ to

$$V^{-1} A^{-1} V^{-\mathsf{T}} = \Lambda^{-1}.$$

Isolating A^{-1} from this equation yields

$$A^{-1} = V\Lambda^{-1} V^{\mathsf{T}} = BB^{\mathsf{T}}.$$

The desired search direction at the point $\bar{x}$ is therefore $-A^{-1} \nabla q(\bar{x})$.

For a *not* necessarily convex-quadratic function $f \in C^1(\mathbb{R}^n, \mathbb{R})$, it remains to be clarified how to introduce a coordinate system at a point $x \in \mathbb{R}^n$ in which f has 'as spherical level sets as possible'. If we limit ourselves to using an approximate construction similar to the one for convex-quadratic functions, this motivates the following definition.

Definition 2.2.29 (Gradient with Respect to a Positive Definite Matrix)
For $f \in C^1(\mathbb{R}^n, \mathbb{R})$ and an (n, n)-matrix $A = A^{\mathsf{T}} \succ 0$,

$$\nabla_A f(x) := A^{-1} \nabla f(x)$$

is called *gradient of f with respect to A* at x.

The various variable metric methods differ in the choice of the matrix A, with the help of which the search direction $-\nabla_A f(x)$ is formed. For each $A = A^\mathsf{T} \succ 0$, this search direction is a first-order descent direction at every noncritical point x, because with A also A^{-1} is positive definite, and it holds

$$\langle \nabla f(\bar{x}), -\nabla_A f(\bar{x}) \rangle = -\nabla f(\bar{x})^\mathsf{T} A^{-1} \nabla f(\bar{x}) < 0.$$

To motivate the term 'variable metric', we first note that every matrix $A = A^\mathsf{T} \succ 0$ defines an inner product as well as a norm.

Exercise 2.2.30 Show that for each $A = A^\mathsf{T} \succ 0$ the function $\langle x, y \rangle_A := x^\mathsf{T} A y$ is an inner product on $\mathbb{R}^n$.

Exercise 2.2.31 Show that, for each $A = A^\mathsf{T} \succ 0$ and the inner product $\langle \cdot, \cdot \rangle_A$ induced by A, the function

$$\|x\|_A := \sqrt{\langle x, x \rangle_A}$$

is a norm on $\mathbb{R}^n$.

The inner product induced by A and the norm induced by A allow further insights.

Exercise 2.2.32 Show that under the assumptions of Theorem 2.2.25 for all $x \in \mathbb{R}^n$

$$\tfrac{1}{2}\|x - x^\star\|_A^2 = q(x) - q(x^\star)$$

holds.

Due to Exercise 2.2.32, Theorem 2.2.25 also makes a statement about the convergence speed of the iterates (and not just of their function values) for the gradient method, provided the appropriate norm is chosen.

Exercise 2.2.33 Show that under the assumptions of Theorem 2.2.25 for all $k \in \mathbb{N}$

$$\|x^{k+1} - x^\star\|_A \leq \frac{\lambda_{\max} - \lambda_{\min}}{\lambda_{\max} + \lambda_{\min}} \|x^k - x^\star\|_A$$

holds.

Exercise 2.2.34 Under the assumptions of Example 2.2.22, show the formula

$$t_e = \frac{\|\nabla q(x)\|_2^2}{\|\nabla q(x)\|_A^2}$$

for the exact step size of the gradient method.

Exercise 2.2.35 For the inner product $\langle \cdot, \cdot \rangle_A$ induced by $A = A^\mathsf{T} \succ 0$ and the induced norm $\| \cdot \|_A$, show the Cauchy-Schwarz inequality

$$\forall\, x, y \in \mathbb{R}^n : \quad |\langle x, y \rangle_A| \le \|x\|_A \cdot \|y\|_A,$$

and also show that the upper bound is attained.

Since the level sets $\{x \in \mathbb{R}^n | \|x\|_A \le c\}$ with $c > 0$ are ellipsoids, the distances around $\bar{x}$ in the norm $\| \cdot \|_A$ are weighted differently, depending on the direction. The following lemma shows that with respect to these new distances, the vector $-\nabla_A f(x)$ is a 'direction of steepest descent'.

Lemma 2.2.36 *Let* $\nabla f(x) \ne 0$. *Then*

$$d = -\nabla_A f(x) \,/\, \|\nabla_A f(x)\|_A$$

is a minimal point of the problem

$$\min \langle \nabla f(x), d \rangle \quad \text{s.t.} \quad \|d\|_A = 1$$

with minimal value $-\|\nabla_A f(x)\|_A$.

Proof According to Exercise 2.2.35, the Cauchy-Schwarz inequality holds for the inner product $\langle \cdot, \cdot \rangle_A$, so for all d with $\|d\|_A = 1$ we obtain

$$\langle \nabla f(x), d \rangle = \langle \nabla_A f(x), d \rangle_A \ge -\|\nabla_A f(x)\|_A \cdot \|d\|_A = -\|\nabla_A f(x)\|_A,$$

where this estimate is sharp. From this follow the claims. □

With the help of any norm $\| \cdot \|$ on $\mathbb{R}^n$, a *metric* on $\mathbb{R}^n$ can be introduced by defining the distance between two points x and y as $\|x - y\|$. We will not use this metric explicitly, but it explains the terminology used: methods that choose a new matrix $A^k = (A^k)^\mathsf{T} \succ 0$ in each iteration and thus define the search direction $-\nabla_{A^k} f(x^k)$ are called *variable metric methods*. Algorithm 2.6 implements this idea.

In the input of Algorithm 2.6, one often chooses $A^0 = I$, so the first search direction is the gradient direction $d^0 = -\nabla f(x^0)$. In line 3, a more consistent termination criterion seems to be $\|\nabla_{A^k} f(x^k)\|_{A^k} \le \varepsilon$, but because of

$$\|\nabla_{A^k} f(x^k)\|_{A^k} = \sqrt{Df(x^k)(A^k)^{-1} \nabla f(x^k)} = \|\nabla f(x^k)\|_{(A^k)^{-1}}$$

Algorithm 2.6: Variable metric method

Input: C^1-optimization problem P, starting point x^0, starting matrix $A^0 = (A^0)^\mathsf{T} \succ 0$
  and termination tolerance $\varepsilon > 0$
Output: Approximation $\bar{x}$ of a critical point of f (if the method terminates; Theorem 2.2.40)

```
 1 begin
 2     Set k = 0.
 3     while ‖∇f(x^k)‖_2 > ε do
 4         Set d^k = −∇_{A^k} f(x^k).
 5         Determine a step size t^k.
 6         Set x^{k+1} = x^k + t^k d^k.
 7         Choose A^{k+1} = (A^{k+1})^⊤ ≻ 0.
 8         Replace k with k + 1.
 9     end
10     Set x̄ = x^k.
11 end
```

---

and the equivalence of $\| \cdot \|_{(A^k)^{-1}}$ and $\| \cdot \|_2$ (i.e., there exist constants $c_1, c_2 > 0$ such that all $x \in \mathbb{R}^n$ satisfy the estimates $c_1 \|x\|_{(A^k)^{-1}} \leq \|x\|_2 \leq c_2 \|x\|_{(A^k)^{-1}}$ [17]), one can just as well test the given and less intricate criterion.

In implementations of line 4, the search direction d^k is usually not calculated by the definition $-(A^k)^{-1}\nabla f(x^k)$ containing a matrix inversion, but less elaborate as the solution of the linear system of equations $A^k d = -\nabla f(x^k)$.

If one wants to guarantee the convergence of variable metric methods in the sense of Theorem 2.2.9, in addition to efficient step sizes one also needs gradient-related search directions. This is not automatic in variable metric methods, but must be required.

Definition 2.2.37 (Uniformly Positive Definite and Bounded Matrices)
A sequence (A^k) of symmetric (n, n)-matrices is called *uniformly positive definite* and *bounded*, if

$$\exists \, 0 < c_1 \leq c_2 \; \forall \, d \in B_=(0, 1), \; k \in \mathbb{N}: \quad c_1 \leq d^\mathsf{T} A^k d \leq c_2$$

holds.

For example, the matrices

$$A^k = \begin{pmatrix} k & 0 \\ 0 & \frac{1}{k} \end{pmatrix}, \quad k \in \mathbb{N},$$

form a sequence that is neither uniformly positive definite nor bounded.

Exercise 2.2.38 Let the sequence (A^k) be uniformly positive definite and bounded with constants c_1 and c_2, respectively. Show that then the sequence $((A^k)^{-1})$ is uniformly positive definite and bounded with constants $1/c_2$ and $1/c_1$, respectively. Also show that the sequence $(\lambda_{\max}((A^k)^{-1})$ of the largest eigenvalues of $((A^k)^{-1})$ is bounded above by $1/c_1$.

Theorem 2.2.39 *Let the sequence (A^k) be uniformly positive definite and bounded. Then the sequence (d^k) with $d^k = -(A^k)^{-1}\nabla f(x^k)$, $k \in \mathbb{N}$, is gradient-related.*

Proof Let $k \in \mathbb{N}$. In the case $\nabla f(x^k) = 0$ there is nothing to show, therefore in the following let $\nabla f(x^k) \neq 0$. By Exercise 2.2.38 it holds firstly

$$\langle \nabla f(x^k), d^k \rangle = -Df(x^k)(A^k)^{-1}\nabla f(x^k) \leq -\frac{1}{c_2}\|\nabla f(x^k)\|_2^2$$

and, secondly, by (2.10)

$$\|d^k\|_2 \leq \|(A^k)^{-1}\|_2 \cdot \|\nabla f(x^k)\|_2 \leq \frac{1}{c_1}\|\nabla f(x^k)\|_2,$$

where we have used Exercise 2.2.38 to upper estimate the spectral norm $\|A\|_2 = \lambda_{\max}(A)$ for a symmetric positive definite matrix A (Remark 2.2.41). Since the numerator of the first fraction in the following estimate is negative, it follows for all $k \in \mathbb{N}$

$$\frac{\langle \nabla f(x^k), d^k \rangle}{\|d^k\|_2} \leq -\frac{c_1}{c_2}\frac{\|\nabla f(x^k)\|_2^2}{\|\nabla f(x^k)\|_2} = -c \cdot \|\nabla f(x^k)\|_2.$$

$\square$

Theorem 2.2.40 *Let the set $f_{\leq}^{f(x^0)}$ be bounded, let the function ∇f be Lipschitz continuous on $\mathrm{conv}(f_{\leq}^{f(x^0)})$, let the sequence (A^k) be uniformly positive definite and bounded, and in line 5 let exact step sizes (t_e^k) or Armijo step sizes (t_a^k) be chosen. Then Algorithm 2.6 terminates after a finite number of steps.*

Proof The claim follows from Corollary 2.2.10 and Theorems 2.2.39, 2.2.16, and 2.2.18. $\square$

Remark 2.2.41 (Spectral Norm and Eigenvalues) Since the spectral norm of a matrix A introduced in Sect. 2.2.3,

$$\|A\|_2 := \max\{\|Ad\|_2 \mid \|d\|_2 = 1\},$$

is the maximal value of a constrained nonlinear optimization problem, the techniques from Chap. 3 will allow us to derive the explicit formula

$$\|A\|_2 = \sqrt{\lambda_{\max}(A^\mathsf{T} A)},$$

where $\lambda_{\max}(A^\mathsf{T} A)$ denotes the largest eigenvalue of the positive semidefinite matrix $A^\mathsf{T} A$ (Example 3.2.46). The fact that the set of eigenvalues of a matrix A is sometimes referred to as its *spectrum* explains the terminology for $\|A\|_2$.

In the case $A = A^\mathsf{T}$, the eigenvalues of $A^\mathsf{T} A = A^2$ are just the squared eigenvalues of A. Then it follows

$$\|A\|_2 = \sqrt{\lambda_{\max}(A^\mathsf{T} A)} = \sqrt{\lambda_{\max}(A^2)} = \sqrt{\max\{\lambda^2 \mid \lambda \text{ eigenvalue of } A\}}$$

$$= \max\{|\lambda| \mid \lambda \text{ eigenvalue of } A\}$$

and in the case $A \succ 0$ also the above formula $\|A\|_2 = \lambda_{\max}(A)$.

With the techniques explained in this section, the explicit formula for $\|A\|_2$ for matrices A of full column rank can at least be motivated. Then the matrix $A^\mathsf{T} A$ is indeed positive definite, and we can define the matrix

$$(A^\mathsf{T} A)^{-\frac{1}{2}} := V\Lambda^{-\frac{1}{2}} V^\mathsf{T}$$

with the help of the matrix Λ of eigenvalues of $A^\mathsf{T} A$ and the matrix V of the associated eigenvectors. The substitution $d = (A^\mathsf{T} A)^{-\frac{1}{2}}\eta$ with $\eta \in \mathbb{R}^n$ then leads to

$$\|Ad\|_2^2 = d^\mathsf{T} A^\mathsf{T} A d = \eta^\mathsf{T}(A^\mathsf{T} A)^{-\frac{1}{2}}(A^\mathsf{T} A)(A^\mathsf{T} A)^{-\frac{1}{2}}\eta = \|\eta\|_2^2.$$

Because of

$$\|A\|_2 = \max\{\|Ad\|_2 \mid \|d\|_2 = 1\} = \max\{\|\eta\|_2 \mid \|(A^\mathsf{T} A)^{-\frac{1}{2}}\eta\|_2 = 1\}$$

the spectral norm of A corresponds to the 'maximal distortion' of the ellipsoid

$$\{\eta \in \mathbb{R}^n \mid \eta^\mathsf{T}(A^\mathsf{T} A)^{-\frac{1}{2}}(A^\mathsf{T} A)^{-\frac{1}{2}}\eta = 1\} = \{\eta \in \mathbb{R}^n \mid \eta^\mathsf{T}(A^\mathsf{T} A)^{-1}\eta = 1\}.$$

As seen in Remark 2.2.26 and Exercise 2.2.27, the longest *semi-axis* of this ellipsoid has the length

$$\frac{1}{\sqrt{\lambda_{\min}((A^\mathsf{T} A)^{-1})}} = \sqrt{\lambda_{\max}(A^\mathsf{T} A)},$$

which corresponds to the claimed value of $\|A\|_2$ and where we have used that the eigenvalues of the positive definite matrix $(A^\mathsf{T} A)^{-1}$ are exactly the reciprocals of the eigenvalues of $A^\mathsf{T} A$. At least for $n = 2$ and $n = 3$ it is geometrically clear that the greatest distortion of an ellipsoid actually occurs along the longest semi-axis, but this is not yet a proof, but only a motivation. A proof will be given in Example 3.2.46.

2.2.5   Newton's Method with and Without Damping

If, for $f \in C^2(\mathbb{R}^n, \mathbb{R})$, one chooses in each iteration of Algorithm 2.6 the matrix $A^k = D^2 f(x^k)$, and if this matrix is positive definite, then one obtains the Newton method which we already briefly mentioned at the beginning of Sect. 2.2.1. However, the Newton steps are 'damped' by the factor t^k, which usually lies in the interval $(0, 1)$. This is sometimes referred to as the *damped Newton method*. We have already pointed out that Newton's method is only well-defined and fast for x^0 sufficiently close to a solution $x^\star$. More precisely, the following applies.

If $x^\star$ is a nondegenerate local minimal point of f, then due to the continuity reasons mentioned at the end of Sect. 2.1.5, $D^2 f(x) \succ 0$ holds for all x in some neighborhood of $x^\star$. For x^0 from this neighborhood, one can thus set $A^k = D^2 f(x^k)$ and obtains a well-defined descent method. Furthermore, the search directions $d^k = -(D^2 f(x^k))^{-1} \nabla f(x^k)$ are gradient-related if f is strongly convex around $x^\star$, i.e., if for some neighborhood U of $x^\star$

$$\exists\, c > 0 \,\forall\, x \in U, \ d \in B_=(0, 1): \quad c \le d^{\mathsf{T}} D^2 f(x) d$$

holds [36]. The boundedness of the sequence $(D^2 f(x^k))$, which is additionally required to conclude this from Theorem 2.2.39, results from the continuity of $D^2 f$. The assumed nondegeneracy of the local minimal point $x^\star$ is automatic for strongly convex f.

The damping of the Newton method has the advantage that the radius of convergence (i.e., the possible distance from x^0 to $x^\star$) may become larger. On the other hand, so far it is not clear whether damping may also slow down the local convergence speed. In any case, the *un*damped Newton method converges quadratically under mild assumptions.

Theorem 2.2.42 (Quadratic Convergence of Newton's Method) *Let the sequence (x^k) defined by*

$$x^{k+1} = x^k - (D^2 f(x^k))^{-1} \nabla f(x^k)$$

converge to a nondegenerate local minimal point $x^\star$, and let $D^2 f$ be Lipschitz continuous on a convex neighborhood of $x^\star$. Then (x^k) converges quadratically to $x^\star$.

Proof First, the definition of the Newton method yields

$$\|x^{k+1} - x^\star\|_2 = \|x^k - (D^2 f(x^k))^{-1} \nabla f(x^k) - x^\star\|_2. \tag{2.11}$$

The proof of quadratic convergence is based on a first-order Taylor expansion of $\nabla f(x^k)$ around $x^\star$ in (2.11) and the quadratic estimation of the resulting error term using Lemma 2.2.13. Since in this textbook we have so far only considered Taylor expansions of real-valued (and not vector-valued) functions, we proceed as follows: According to Theorem 2.1.30a for each $i \in \{1, \ldots, n\}$ the error term

$$w_i := \partial_{x_i} f(x^k) - \partial_{x_i} f(x^\star) - \langle \nabla \partial_{x_i} f(x^\star), x^k - x^\star \rangle$$

satisfies

$$w_i = o(\|x^k - x^\star\|).$$

In addition, the function $\nabla \partial_{x_i} f$ is Lipschitz continuous with constant $L > 0$ on the convex neighborhood D of $x^\star$, on which also the Lipschitz continuity of $D^2 f$ with constant $L > 0$ is assumed, so the condition

$$\forall x, y \in D : \quad \|D^2 f(x) - D^2 f(y)\|_2 \leq L \|x - y\|_2$$

is fulfilled. Indeed, due to $\|A\|_2 = \max_{\|d\|_2=1} \|Ad\|_2$ and $\|e_i\|_2 = 1$ for all $x, y \in D$ one obtains

$$\|\nabla \partial_{x_i} f(x) - \nabla \partial_{x_i} f(y)\|_2 = \|(D^2 f(x) - D^2 f(y)) e_i\|_2 \leq \|D^2 f(x) - D^2 f(y)\|_2$$
$$\leq L \|x - y\|_2.$$

For all sufficiently large $k \in \mathbb{N}$, the iterates x^k lie in D, so that Lemma 2.2.13 yields for the error term even

$$|w_i| \leq \frac{L}{2} \|x^k - x^\star\|_2^2.$$

With the vector of all error terms

$$w = \nabla f(x^k) - \nabla f(x^\star) - D^2 f(x^\star)(x^k - x^\star) = \nabla f(x^k) - D^2 f(x^\star)(x^k - x^\star)$$

we can replace $\nabla f(x^k)$ in (2.11) and derive an upper bound. We also use that, due to continuity reasons, $\|(D^2 f(x^k))^{-1}\|_2$ is bounded by a constant $c > 0$ in some neighborhood of $x^\star$.

$$\|x^{k+1} - x^\star\|_2 = \|x^k - (D^2 f(x^k))^{-1} \nabla f(x^k) - x^\star\|_2$$
$$= \|x^k - x^\star - (D^2 f(x^k))^{-1} \left(D^2 f(x^\star)(x^k - x^\star) + w \right)\|_2$$
$$= \|(D^2 f(x^k))^{-1} \left((D^2 f(x^k) - D^2 f(x^\star))(x^k - x^\star) + w \right)\|_2$$
$$\leq \|(D^2 f(x^k))^{-1}\|_2 \left(\|(D^2 f(x^k) - D^2 f(x^\star))(x^k - x^\star)\|_2 + \|w\|_2 \right)$$
$$\leq c \left(\|D^2 f(x^k) - D^2 f(x^\star)\|_2 \|x^k - x^\star\|_2 + \|w\|_2 \right)$$
$$\leq c \left(L \|x^k - x^\star\|_2^2 + \frac{L\sqrt{n}}{2} \|x^k - x^\star\|_2^2 \right)$$
$$= c \left(L + \frac{L\sqrt{n}}{2} \right) \|x^k - x^\star\|_2^2.$$

This proves the claim. $\qquad \square$

Remark 2.2.43 The assumptions of Theorem 2.2.42 can be weakened considerably. First, the statement holds for every nondegenerate critical point $x^\star$, not just for nondegenerate local minimal points. Second, we will see in Theorem 2.2.52 that the convergence of the sequence (x^k) already implies that the limit point $x^\star$ is a nondegenerate critical point.

The convergence speed from Theorem 2.2.42 transfers to the damped Newton method, if one chooses $t^k = 1$ for all sufficiently large $k \in \mathbb{N}$. The following exercise provides a natural condition for this.

Exercise 2.2.44 For $f \in C^2(\mathbb{R}^n, \mathbb{R})$, let x lie in a sufficiently small neighborhood of a nondegenerate local minimal point, and let the search direction d be determined by the damped Newton method, using the Armijo rule with $t^0 = 1$ and $\sigma < \frac{1}{2}$. Show that then $\langle \nabla f(x), d \rangle < 0$ holds and that the Armijo rule selects the step size $t_a = 1$.

Exercise 2.2.45 Show that for the function $q(x) = \frac{1}{2} x^\mathsf{T} A x + b^\mathsf{T} x$ with $A = A^\mathsf{T} \succ 0$ and $b \in \mathbb{R}^n$, and for every starting point $x^0 \in \mathbb{R}^n$, the undamped Newton method generates the global minimal point of q in one iteration.

Exercise 2.2.46 For $f \in C^2(\mathbb{R}^n, \mathbb{R})$ let an iterate x^k with $D^2 f(x^k) \succ 0$ be given. Show that then the search direction d^k generated by Newton's method is the unique global minimal point of the convex-quadratic function

$$q^k(d) = f(x^k) + \langle \nabla f(x^k), d \rangle + \tfrac{1}{2} d^\mathsf{T} D^2 f(x^k) d.$$

According to Exercise 2.2.46, the basic idea of Newton's method, namely to compute a zero of ∇f by iteratively finding *zeros of linear approximations* to ∇f, can therefore also be interpreted as iteratively computing *minimal points of quadratic approximations* to f, in order to minimize f. We will return to this interpretation when considering conjugate gradient, trust-region and sequential quadratic programming methods.

We briefly mention an important modification of Newton's method for objective functions f with the special structure of *least squares problems* like in Example 1.1.4 and Exercise 2.1.18, i.e., $f(x) = \frac{1}{2} \|r(x)\|_2^2$ with a smooth function $r : \mathbb{R}^n \to \mathbb{R}^m$. If r is linear, one speaks of a *linear* least squares problem (e.g., [25, 26]). Since f is then easily seen to be convex-quadratic (Exercise 2.1.44), algorithmically this case can be treated efficiently [26]. For *nonlinear* least squares problems, the Hessian matrix is calculated by the chain and product rules as

$$D^2 f(x) = \nabla r(x) \, Dr(x) + \sum_{j=1}^m r_j(x) \, D^2 r_j(x). \tag{2.12}$$

The basic idea of the central method for solving nonlinear least squares problems, namely the *Gauss-Newton method*, is to choose $A^k = \nabla r(x^k) \, Dr(x^k)$ in Algorithm 2.6 instead of $A^k = D^2 f(x^k)$, as in Newton's method. The fact that the remaining terms in the representation of $D^2 f(x^k)$ play a subordinate role can either be due to the fact that for $m \leq n$ a point $x^\star$ with $r(x^\star) = 0$ is usually approximated, so that the values $r_j(x^k)$ almost vanish, or to the fact that the curvatures of the functions r_j at $x^\star$ are negligible, so that the matrices $D^2 r_j(x^k)$ are close to the zero matrix.

Although for setting up A^k in the the Gauss-Newton method only first-order derivative information (the matrix $Dr(x^k)$) is required, under certain additional conditions still quadratic convergence can be shown [26]. Moreover, the search directions d^k in the Gauss-Newton method are guaranteed to be descent directions (of first order), so that step size control, for example by the Armijo rule, is possible.

If the Jacobian matrix $Dr(x^k)$ does not possess full rank or if it is poorly conditioned, the Gauss-Newton method can be stabilized by choosing $A^k = \nabla r(x^k) Dr(x^k) + \sigma^k I$ with a certain value $\sigma^k > 0$ and with the identity matrix I of appropriate dimension. This leads to the Levenberg-Marquardt method (Exercise 2.2.49) which can be considered a precursor to the trust-region methods introduced in Sect. 2.2.10 (for details see Example 3.2.25 and [26]).

Exercise 2.2.47 For least squares problems, show the representation of the Hessian matrix $D^2 f(x)$ from (2.12).

Exercise 2.2.48 Let A be a symmetric (n, n)-matrix and I the (n, n)-identity matrix. Show that for all sufficiently large $\sigma \in \mathbb{R}$ the matrix $A + \sigma I$ is positive definite.

Exercise 2.2.49 For $f \in C^2(\mathbb{R}^n, \mathbb{R})$ and an iterate x^k, let the matrix A^k be a symmetric, but not necessarily positive definite, approximation to the Hessian matrix $D^2 f(x^k)$ (which includes the case $A^k = D^2 f(x^k) \not\succ 0$). The Levenberg-Marquardt approach consists in choosing some $\sigma^k > 0$ with $A^k + \sigma^k I \succ 0$ (which exists by Exercise 2.2.48) and defining the search direction $d^k = -(A^k + \sigma^k I)^{-1} \nabla f(x^k)$. Show that d^k is the unique global minimal point of the convex-quadratic function

$$q_{\sigma^k}^k(d) = f(x^k) + \langle \nabla f(x^k), d \rangle + \tfrac{1}{2} d^\mathsf{T} A^k d + \tfrac{\sigma^k}{2} \|d\|_2^2.$$

From the perspective of Exercise 2.2.49, the Levenberg-Marquardt approach regularizes the nonconvex quadratic function $q^k(d) = f(x^k) + \langle \nabla f(x^k), d \rangle + \tfrac{1}{2} d^\mathsf{T} A^k d$ by the strongly convex term $\|d\|_2^2/2$ with regularization parameter σ^k. This technique is known as *Tikhonov regularization*.

2.2.6 Superlinear Convergence

If in Newton's method x^0 is not sufficiently close to a nondegenerate minimal point, then $D^2 f(x^k)$ is not necessarily positive definite and the Newton direction $d^k = -(D^2 f(x^k))^{-1} \nabla f(x^k)$ is either not defined or not necessarily a descent direction.

Therefore, one tries to *globalize* the Newton method, i.e., to enforce convergence in the sense of Theorem 2.2.9 towards a local minimal point from *every* starting point $x^0 \in \mathbb{R}^n$. This should not be confused with convergence towards a *global* minimal point (for such methods see [36]).

To this end, a first approach chooses $A^0 = I$ as the input for Algorithm 2.6, and in line 7

$$A^{k+1} \;=\; D^2 f(x^{k+1}) + \sigma^{k+1} I$$

with a scalar σ^{k+1} so large that A^{k+1} is positive definite (Exercises 2.2.48 and 2.2.49).

Then $d^0 \;=\; -\nabla f(x^0)$ holds, and if convergence towards a nondegenerate local minimal point occurs, one can choose $\sigma^k = 0$ for all sufficiently large k (i.e., the method starts as the gradient method, and after a finite number of iterations it transitions into the damped Newton method). Under suitable conditions, one can show superlinear convergence of the method (e.g., by Theorem 2.2.52). A disadvantage of the method is that determining σ^k can be costly: For example, one halves or doubles σ^k until a test for positive definiteness of $D^2 f(x^k) + \sigma^k I$ is successful. The method is therefore not used in this form in practice.

In the following, we will discuss methods that do not transition into the damped Newton method after *finitely many* iterations, but only *asymptotically*. Also for these, at least superlinear convergence can be shown. The corresponding convergence theorem requires some preparation.

First, the sequence of iterates (x^k) has an accumulation point according to Theorem 2.2.9, and every such accumulation point is critical, provided the set $f_{\leq}^{f(x^0)}$ is bounded and gradient-related search directions and efficient step sizes are used. The gradient-relatedness of the search directions is guaranteed by Theorem 2.2.39 for uniformly positive definite and bounded (A^k). With the means of this textbook we can only show in simple cases that the sequence (x^k) actually *converges* (e.g., if f has a unique critical point), but we will assume it for the subsequent investigations of the convergence *speed*.

For abbreviation, we set

$$H^k \;:=\; t^k \left(A^k \right)^{-1},$$

i.e., in line 6 of algorithm 2.6 the new iterate is chosen as

$$x^{k+1} \;=\; x^k - H^k \nabla f(x^k). \tag{2.13}$$

We will also use that the definition of superlinear convergence of a sequence (x^k) converging to $x^\star$ (Definition 2.2.23b) is equivalent to

$$\limsup_k \frac{\|x^{k+1} - x^\star\|}{\|x^k - x^\star\|} \;=\; 0.$$

Lemma 2.2.50 *Let the sequence (x^k) be formed according to the rule (2.13) and convergent to $x^\star$. Furthermore, let the sequences $\left(\|H^k\|_2\right)$ and $\left(\|\left(H^k\right)^{-1}\|_2\right)$ be bounded. Then the following holds:*

(a) $\nabla f(x^\star) = 0$.
(b) $\limsup_k \|x^{k+1} - x^\star\|_2/\|x^k - x^\star\|_2 \le \limsup_k \|I - H^k D^2 f(x^\star)\|_2$.

Proof Because of

$$0 = x^\star - x^\star = \lim_k (x^k - x^{k+1}) = \lim_k H^k \nabla f(x^k)$$

and the boundedness of $\left(\|(H^k)^{-1}\|_2\right)$, it follows

$$\|\nabla f(x^k)\|_2 = \|(H^k)^{-1} H^k \nabla f(x^k)\|_2 \le \|(H^k)^{-1}\|_2 \cdot \|H^k \nabla f(x^k)\|_2 \to 0$$

and thus also

$$\|\nabla f(x^\star)\|_2 = \lim_k \|\nabla f(x^k)\|_2 = 0,$$

hence statement a.

With $z(s) = x^\star + s(x^k - x^\star)$ this implies

$$\|x^{k+1} - x^\star\|_2 = \|x^k - x^\star - H^k \nabla f(x^k)\|_2$$

$$= \left\| x^k - x^\star - H^k \int_0^1 D^2 f(z(s))(x^k - x^\star)\, ds \right\|_2$$

$$= \left\| x^k - x^\star - H^k \left(D^2 f(x^\star)(x^k - x^\star) + \int_0^1 \left(D^2 f(z(s)) - D^2 f(x^\star)\right)(x^k - x^\star)\, ds \right) \right\|_2$$

$$\le \|I - H^k D^2 f(x^\star)\|_2 \cdot \|x^k - x^\star\|_2 + \|H^k\|_2 \cdot \left\| \int_0^1 \left(D^2 f(z(s)) - D^2 f(x^\star)\right)(x^k - x^\star)\, ds \right\|_2.$$

It follows

$$\frac{\|x^{k+1} - x^\star\|_2}{\|x^k - x^\star\|_2}$$

$$\le \|I - H^k D^2 f(x^\star)\|_2 + \|H^k\|_2 \cdot \left\| \int_0^1 \left(D^2 f(z(s)) - D^2 f(x^\star)\right) \frac{x^k - x^\star}{\|x^k - x^\star\|_2}\, ds \right\|_2.$$

Since the sequences $\left(\|H^k\|_2\right)$ and $\left(x^k - x^\star/\|x^k - x^\star\|_2\right)$ are bounded and

$$D^2 f(z(s)) - D^2 f(x^\star) = D^2 f(x^\star + s(x^k - x^\star)) - D^2 f(x^\star)$$

converges to the zero matrix for each $s \in [0, 1]$, statement b is shown. □

Lemma 2.2.51 *For two (n, n)-matrices A and B, let $L := \|I - AB\|_2 < 1$. Then the following holds:*

(a) *A and B are nonsingular.*
(b) *$\|A\|_2 \leq (1 + L) \cdot \|B^{-1}\|_2$.*
(c) *$\|A^{-1}\|_2 \leq \|B\|_2/(1 - L)$.*

Proof Because of $\|I - AB\|_2 < 1$ the absolute values of the eigenvalues of $I - AB$ are strictly less than one (Example 3.2.46). Therefore, the matrix AB cannot have the eigenvalue zero, so it is nonsingular. Consequently, A and B are also nonsingular, and statement a is shown.

Statement b follows from

$$\|A\|_2 = \|ABB^{-1}\|_2 \leq \|AB\|_2 \cdot \|B^{-1}\|_2 = \|I - (I - AB)\|_2 \cdot \|B^{-1}\|_2$$

$$\leq (\|I\|_2 + \|I - AB\|_2) \cdot \|B^{-1}\|_2 = (1 + L) \cdot \|B^{-1}\|_2.$$

Finally, let $C := AB$ and $z \in B_=(0, 1)$ be a vector with $\|C^{-1}z\|_2 = \|C^{-1}\|_2$. Then with $u := C^{-1}z$ and $v := (I - C)u = u - z$

$$\|v\|_2 \leq \|I - C\|_2 \cdot \|u\|_2 = L \cdot \|C^{-1}\|_2$$

holds, and therefore

$$\|C^{-1}\|_2 = \|u\|_2 = \|v + z\|_2 \leq \|v\|_2 + \|z\|_2 \leq L \cdot \|C^{-1}\|_2 + 1.$$

It follows

$$\|B^{-1}A^{-1}\|_2 = \|C^{-1}\|_2 \leq \frac{1}{1 - L},$$

so

$$\|A^{-1}\|_2 = \|BB^{-1}A^{-1}\|_2 \leq \|B\|_2 \cdot \|B^{-1}A^{-1}\|_2 \leq \frac{\|B\|_2}{1 - L}$$

and thus statement c.   □

Theorem 2.2.52 *Let the sequence (x^k) be formed according to the rule (2.13) and convergent to $x^\star$. Furthermore, let $L := \limsup_k \|I - H^k D^2 f(x^\star)\|_2 < 1$. Then the following statements hold:*

(a) *$D^2 f(x^\star)$ is nonsingular.*
(b) *$\nabla f(x^\star) = 0$.*
(c) *(x^k) converges at least linearly to $x^\star$.*
(d) *$L = 0$ holds exactly in the case $\lim_k H^k = \left(D^2 f(x^\star)\right)^{-1}$, and in this case (x^k) converges superlinearly to $x^\star$.*

Proof In view of $\limsup_k \|I - H^k D^2 f(x^\star)\|_2 < 1$, there exists some $k_0 \in \mathbb{N}$, such that for all $k \geq k_0$

$$\|I - H^k D^2 f(x^\star)\|_2 < 1$$

is fulfilled. According to Lemma 2.2.51a, $D^2 f(x^\star)$ is therefore nonsingular, which proves statement a, and moreover, for $k \geq k_0$, H^k is also nonsingular. From Lemma 2.2.51b and c, the boundedness of the sequences $\left(\|H^k\|_2\right)$ and $\left(\|\left(H^k\right)^{-1}\|_2\right)$ follows, so that Lemma 2.2.50a provides statement b. According to Lemma 2.2.50b, we also have

$$\limsup_k \frac{\|x^{k+1} - x^\star\|_2}{\|x^k - x^\star\|_2} \leq L,$$

which proves statement c. For $L = 0$ this implies superlinear convergence.

The rest of statement d is shown as follows. With $\bar{H} := (D^2 f(x^\star))^{-1}$, it holds

$$\|\bar{H} - H^k\|_2 = \|(I - H^k \bar{H}^{-1})\bar{H}\|_2 \leq \|I - H^k \bar{H}^{-1}\|_2 \cdot \|\bar{H}\|_2$$

and

$$\|I - H^k D^2 f(x^\star)\|_2 = \|(\bar{H} - H^k)D^2 f(x^\star)\|_2 \leq \|\bar{H} - H^k\|_2 \cdot \|D^2 f(x^\star)\|_2.$$

With this, we obtain

$$\limsup_k \|I - H^k D^2 f(x^\star)\|_2 = 0$$

exactly for

$$\lim_k H^k = \bar{H} = (D^2 f(x^\star))^{-1}.$$

$\square$

According to Theorem 2.2.52, Algorithm 2.6 should therefore *asymptotically* transition into the undamped Newton method to guarantee superlinear convergence. Because of

$$H^k = t^k \cdot (A^k)^{-1}$$

natural conditions for this are $\lim_k t^k = 1$ and $\lim_k A^k = D^2 f(x^\star)$. The Levenberg-Marquardt approach proposed at the beginning of this section achieves this with high effort already after a finite number of steps, so it is not efficient in this sense.

2.2.7   Quasi-Newton Methods

An alternative idea for the construction of matrices A^k with $\lim_k A^k = D^2 f(x^\star)$ stems from the *secant method* for finding a zero of a function from $\mathbb{R}^1$ to $\mathbb{R}^1$. In the optimization framework, let this be the function $\nabla f = f'$ for a function $f : \mathbb{R} \to \mathbb{R}$

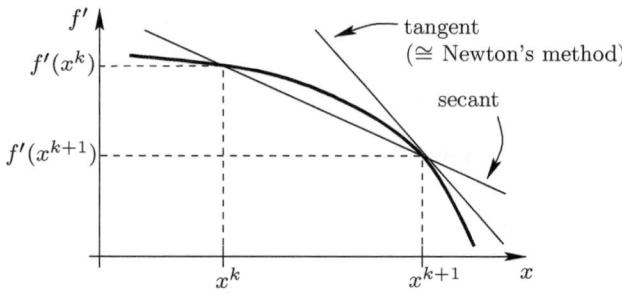

Fig. 2.8 Basic idea of the secant method

which is to be minimized. The secant method differs from the Newton method in
that the linear approximations to f', whose zeros are calculated, are not given by
tangents to f', but by secants. More precisely, the tangent to f' at an iteration point
x^{k+1} is approximated by the secant to the function graph through the two points
$(x^k, f'(x^k))$ and $(x^{k+1}, f'(x^{k+1}))$ (Fig. 2.8).

The resulting line has the slope

$$a^{k+1} = \frac{f'(x^{k+1}) - f'(x^k)}{x^{k+1} - x^k},$$

and the sequence of these secant slopes seems to approximate the tangent slope at
the solution point $x^\star$, that is, $a^k \to f''(x^\star)$. This is the desired sufficient condition
for superlinear convergence from the previous section. Analogously, for $n \geq 1$ the
equation

$$\nabla f(x^{k+1}) - \nabla f(x^k) = A^{k+1} \cdot (x^{k+1} - x^k) \tag{2.14}$$

is referred to as *secant equation* or *quasi-Newton condition* for the symmetric (n, n)-
matrix A^{k+1}. One counts that (2.14) provides n equations for the n^2 entries of
A^{k+1}. In fact, by the symmetry of A^{k+1} only $n(n + 1)/2$ different entries are to be
determined, but for $n > 1$ the n equations then still do not specify A^{k+1} uniquely.
For this reason, there are many possibilities to formulate different quasi-Newton
methods.

The basic idea of the following methods is to not completely recalculate the
matrix A^{k+1} from scratch in each iteration, but to obtain it as a possibly simple
update of the matrix A^k from the previous step. The following approach has turned
out to be successful. One starts with $A^0 = (A^0)^\mathsf{T} \succ 0$, and in line 7 of Algorithm 2.6
the matrix A^{k+1} is derived from A^k by adding a symmetric matrix of rank one or
two:

$$A^{k+1} = A^k + \alpha_k (u^k)(u^k)^\mathsf{T} + \beta_k (v^k)(v^k)^\mathsf{T}$$

with scalars α_k, $\beta_k \in \mathbb{R}$ and vectors u^k, $v^k \in \mathbb{R}^n$, which are chosen such that A^{k+1} satisfies the secant equation (2.14).

The matrices that occur are of the type ab^T with two column vectors a and b. Such a matrix has at most rank one and is called *dyadic product* of the vectors a and b. In our application we additionally put $b := a$, so that the dyadic product is also a symmetric matrix.

We will see that a pure rank-1-update does not necessarily inherit the positive definiteness of the matrix A^k to A^{k+1}, while rank-2-updates can be arranged in this way.

With the abbreviations

$$s^k := x^{k+1} - x^k \qquad \text{and} \qquad y^k := \nabla f(x^{k+1}) - \nabla f(x^k)$$

the secant equation (2.14) for the thus defined matrix A^{k+1} reads

$$y^k = \left(A^k + \alpha_k (u^k)(u^k)^\mathsf{T} + \beta_k (v^k)(v^k)^\mathsf{T} \right) \cdot s^k.$$

In the following, let $k \in \mathbb{N}$ be fixed and omitted. Then it follows

$$y - As = (\alpha \cdot u^\mathsf{T} s) \cdot u + (\beta \cdot v^\mathsf{T} s) \cdot v.$$

By comparing the two summands on the left and right sides of this equation, one can arrive at the obvious choices $u := y$ and $v := As$. The appropriate coefficients on the right side are then obtained for the choices

$$\alpha := \frac{1}{y^\mathsf{T} s} \qquad \text{and} \qquad \beta := -\frac{1}{s^\mathsf{T} As},$$

provided the occurring denominators do not vanish. The resulting update formula

$$A^+ = A + \frac{yy^\mathsf{T}}{y^\mathsf{T} s} - \frac{As\, s^\mathsf{T} A}{s^\mathsf{T} As} \tag{2.15}$$

(where A^+ stands for A^{k+1}) is called *BFGS update* (after Broyden, Fletcher, Goldfarb and Shanno, who independently found this formula in 1970 [5, 9, 14, 30]).

Since in line 4 of Algorithm 2.6 the search direction

$$d^k = -\left(A^k \right)^{-1} \nabla f(x^k)$$

is chosen, it would be advantageous to be able to explicitly specify the matrix $\left(A^k \right)^{-1}$. Due to the simple structure of rank-2-updates, this is actually possible, namely with the help of the *Sherman-Morrison-Woodbury formula*.

Exercise 2.2.53 (Sherman-Morrison-Woodbury Formula)

(a) Show for a nonsingular (n, n)-matrix A and vectors $b, c \in \mathbb{R}^n$, that $A + b\,c^\mathsf{T}$ is nonsingular if and only if $1 + c^\mathsf{T} A^{-1} b$ does not vanish.

(b) Prove the *Sherman-Morrison-Woodbury formula* for an (n, n)-matrix A and vectors $b, c \in \mathbb{R}^n$, where A and $A + b\,c^\mathsf{T}$ are nonsingular:

$$(A + b\,c^\mathsf{T})^{-1} = A^{-1} - \frac{A^{-1} b\,c^\mathsf{T} A^{-1}}{1 + c^\mathsf{T} A^{-1} b}.$$

Exercise 2.2.53 provides an update formula for the *inverse* matrices

$$B := A^{-1} \quad \text{and} \quad B^+ := \left(A^+\right)^{-1},$$

namely

$$B_{\mathrm{BFGS}}^+ = B + \frac{s s^\mathsf{T}}{s^\mathsf{T} y} - \frac{B y\, y^\mathsf{T} B}{y^\mathsf{T} B y} + r r^\mathsf{T} \tag{2.16}$$

with

$$r := \sqrt{y^\mathsf{T} B y} \cdot \left(\frac{s}{s^\mathsf{T} y} - \frac{B y}{y^\mathsf{T} B y} \right).$$

With this knowledge of the inverse matrices, in line 4 of Algorithm 2.6 there is no longer the need to solve the linear system of equations $A^k d = -\nabla f(x^k)$ to determine the search direction d^k, but one just defines d^k as the matrix-vector product $-B^k \nabla f(x^k)$. To write the algorithm completely in terms of matrices B^k rather than A^k, in the input one chooses a matrix $B^0 = (B^0)^\mathsf{T} \succ 0$ instead of A^0, and in line 7 one chooses $B^{k+1} = B_{\mathrm{BFGS}}^{k+1}$.

After this conversion of the algorithm from matrices A to their inverses B, one might ask why the derivation of a rank-2 update was not directly performed for the inverses. Indeed, the secant equation $y = As$ is equivalent to $s = By$. If one carries out the analogous construction of an update for B as above for A, one simply obtains the same formula as in (2.15), only with swapped vectors s and y:

$$B_{\mathrm{DFP}}^+ = B + \frac{s s^\mathsf{T}}{s^\mathsf{T} y} - \frac{B y\, y^\mathsf{T} B}{y^\mathsf{T} B y}.$$

This *DFP update* (after Davidon and Fletcher/Powell [6, 10]) differs from the BFGS update only by the absence of the term $r r^\mathsf{T}$. The introduction of an additional parameter $\theta \in \mathbb{R}$ yields the updates of the *Broyden family*

$$B_\theta^+ = B_{\mathrm{DFP}}^+ + \theta \cdot r r^\mathsf{T}.$$

Obviously, $B_0^+ = B_{\text{DFP}}^+$ and $B_1^+ = B_{\text{BFGS}}^+$ hold. For the choice

$$\theta = \frac{s^\mathsf{T} y}{s^\mathsf{T} y - y^\mathsf{T} B y}$$

one calculates

$$B_{\text{SR1}}^+ := B_\theta^+ = B + \frac{(s - By)(s - By)^\mathsf{T}}{(s - By)^\mathsf{T} y},$$

so that the update matrix only has rank one. This is referred to as the *SR1 update* (*SR1 = symmetric rank one*).

Regarding the division by the numbers $s^\mathsf{T} y$ and $y^\mathsf{T} B y$ in the update formulas, the following can be noted. In Lemma 2.2.55 we will show that for $\theta \geq 0$ with B^0 all iterated matrices B^k are positive definite, provided $(s^k)^\mathsf{T} y^k$ is positive. In particular, then $y^k \neq 0$ and $(y^k)^\mathsf{T} B^k y^k > 0$ are satisfied.

The crucial question is therefore whether $(s^k)^\mathsf{T} y^k$ is always positive. At least for convex-quadratic functions $q(x) = \frac{1}{2} x^\mathsf{T} A x + b^\mathsf{T} x$ with $A = A^\mathsf{T} \succ 0$ we have

$$y^k = \nabla q(x^{k+1}) - \nabla q(x^k) = (Ax^{k+1} + b) - (Ax^k + b) = A(x^{k+1} - x^k)$$
$$= As^k$$

(i.e., A fulfills the secant equation in each iteration), from which

$$(s^k)^\mathsf{T} y^k = (s^k)^\mathsf{T} As^k > 0$$

follows. With some more effort, this result can also be shown for locally strongly convex functions f. Regardless of this, $(s^k)^\mathsf{T} y^k$ is also positive when exact step lengths t_e^k are chosen.

Exercise 2.2.54 In the k-th iteration of a quasi-Newton method, let $B^k = (B^k)^\mathsf{T} \succ 0$ and

$$d^k = -B^k \nabla f(x^k),$$
$$x^{k+1} = x^k + t^k d^k,$$
$$y^k = \nabla f(x^{k+1}) - \nabla f(x^k),$$
$$s^k = x^{k+1} - x^k.$$

Show that, when choosing exact step sizes $t^k = t_e^k$, the inequality

$$(y^k)^\mathsf{T} s^k > 0$$

holds.

The positivity of $(s^k)^\mathsf{T} y^k$ can also be guaranteed by a step size control which combines the Armijo condition (2.3) and the so-called *curvature condition*

$$\forall k \in \mathbb{N} : \quad \langle \nabla f(x^k + t^k d^k), d^k \rangle \geq c_2 \cdot \langle \nabla f(x^k), d^k \rangle$$

with some $c_2 \in (c_1, 1)$ and c_1 from (2.3) satisfying $c_1 \in (0, \frac{1}{2})$ [12, 26]. These two conditions are collectively known as the *Wolfe conditions* or *Wolfe-Powell conditions*.

Apparently, the matrices B_θ^+ inherit the symmetry of B for all $\theta \in \mathbb{R}$. Thus, with a symmetric matrix B^0, all iterated matrices B^k are also symmetric. The following result addresses the inheritance of positive definiteness.

Lemma 2.2.55 *Let $\theta \geq 0$ be arbitrary. Then under the conditions $B = B^\mathsf{T} \succ 0$ and $s^\mathsf{T} y > 0$ also $B_\theta^+ \succ 0$ is true.*

Proof For all $w \neq 0$ it holds

$$w^\mathsf{T} B_\theta^+ w = w^\mathsf{T} B w + \frac{(w^\mathsf{T} s)^2}{s^\mathsf{T} y} - \frac{(w^\mathsf{T} B y)^2}{y^\mathsf{T} B y} + \theta \cdot (w^\mathsf{T} r)^2.$$

Since

$$w^\mathsf{T} B w - \frac{(w^\mathsf{T} B y)^2}{y^\mathsf{T} B y} = \|w\|_B^2 - \frac{\langle w, y \rangle_B^2}{\|y\|_B^2}$$

is nonnegative according to the Cauchy-Schwarz inequality, and the same applies to the other two summands, $w^\mathsf{T} B^+ w$ is nonnegative.

It remains to show that $w^\mathsf{T} B^+ w$ is indeed positive. This is guaranteed in the case

$$\|w\|_B^2 - \frac{\langle w, y \rangle_B^2}{\|y\|_B^2} > 0.$$

If this expression vanishes, then according to the Cauchy-Schwarz inequality there exists some $\lambda \neq 0$ with $w = \lambda y$ for $w \neq 0$. In this case we obtain

$$\frac{(w^\mathsf{T} s)^2}{s^\mathsf{T} y} = \lambda^2 s^\mathsf{T} y > 0,$$

so the claim is also valid. □

While the condition $\theta \geq 0$ from Lemma 2.2.55 is not guaranteed for the SR1 update, the latter still can be observed to deliver good results in practice.

If one chooses exact step sizes $t^k = t_e^k$, $k \in \mathbb{N}$, then for any $\theta \in \mathbb{R}$ the search direction can be calculated to be

$$d^{k+1} = -B_\theta^{k+1} \nabla f(x^{k+1}) = \left(\frac{(y^k)^\intercal d^k}{\sqrt{(y^k)^\intercal B^k y^k}} - \theta (r^k)^\intercal \nabla f(x^{k+1}) \right) \cdot r^k.$$

Since only the *coefficient* of the vector r^k depends on θ, the search *direction* is identical for every $\theta \in \mathbb{R}$. Since the one-dimensional minimization along this direction is exact, *all methods of the Broyden family deliver identical solution sequences* (x^k).

This surprising result is put into perspective by the fact that in practice one usually does *not* use exact, but inexact step sizes, computed for example by the Armijo rule with backtracking line search. With inexact step size control, the solution sequences from different methods of the Broyden family can indeed differ quite significantly. While, for example, the DFP update tends to generate poorly conditioned matrices B^k, the BFGS update numerically often behaves robust for problems of medium size. Here, 'medium size' means that the space required to store the matrices B^k is not too large (for example in relation to the size of the computer's main memory).

Unfortunately, the matrices B^k do not always tend towards $(D^2 f(x^\star))^{-1}$, as would be desirable for the application of Theorem 2.2.52 for superlinear convergence. With a rather technical generalization of Theorem 2.2.52 the superlinear convergence of the BFGS and DFP methods can nevertheless be shown if $(B^k)^{-1}$ and $D^2 f(x^\star)$ are at least asymptotically equal *along the search directions* d^k, and if $\lim_k t^k = 1$ holds (for details see [26]).

2.2.8 Conjugate Directions

In many applications the number of variables is so high that vectors of length n like x^k and d^k can still be stored well, but the storage of the $n(n+1)/2$ entries of matrices like B^k is problematic. In order to derive more efficient methods than the gradient method for such high-dimensional problems, in this section we deal with the special role that orthogonality with respect to the inner product $\langle \cdot, \cdot \rangle_A$ plays.

Definition 2.2.56 (Conjugacy with Respect to a Positive Definite Matrix)
Let A be an (n, n)-matrix with $A = A^\intercal \succ 0$. Two vectors $v, w \in \mathbb{R}^n$ are called *conjugate with respect to A*, if $\langle v, w \rangle_A = 0$ holds.

In the following, we consider the general descent method

$$x^{k+1} = x^k + t_e^k d^k$$

with exact step sizes t_e^k and first-order descent directions d^k for the convex-quadratic function

$$q(x) = \tfrac{1}{2} x^\mathsf{T} A x + b^\mathsf{T} x$$

with $A = A^\mathsf{T} \succ 0$ and $b \in \mathbb{R}^n$. The first result will be that search directions conjugate with respect to A lead to a faster identification of a global minimal point than, for example, the negative gradient directions of q.

Exercise 2.2.57 For $k \in \mathbb{N}$, let $d^0, \ldots, d^k$ be pairwise conjugate with respect to A and all nonzero. Show:

(a) The vectors $d^0, \ldots, d^k$ are linearly independent. In particular, we have $k < n$.
(b) For $k = n - 1$ it holds

$$A^{-1} = \sum_{\ell=0}^{n-1} \frac{(d^\ell)(d^\ell)^\mathsf{T}}{(d^\ell)^\mathsf{T} A (d^\ell)}.$$

Lemma 2.2.58 *For $k \in \mathbb{N}$, let $d^0, \ldots, d^k$ be pairwise conjugate with respect to A. Then it holds*

$$\forall\, 0 \leq \ell \leq k: \quad \langle \nabla q(x^{k+1}), d^\ell \rangle = 0.$$

Proof For $\ell = k$, the assertion results from (2.8), because the exact step size t_e^k satisfies

$$0 = \varphi'_{d^k}(t_e^k) = \langle \nabla q(x^k + t_e^k d^k), d^k \rangle = \langle \nabla q(x^{k+1}), d^k \rangle.$$

For all $0 \leq \ell \leq k - 1$, it holds

$$x^{k+1} = x^k + t_e^k d^k = x^{\ell+1} + \sum_{j=\ell+1}^{k} t_e^j d^j,$$

thus

$$\nabla q(x^{k+1}) - \nabla q(x^{\ell+1}) = (A x^{k+1} + b) - (A x^{\ell+1} + b) = \sum_{j=\ell+1}^{k} t_e^j A d^j$$

and therefore, due to $0 = \varphi'_{d^\ell}(t_e^\ell) = \langle \nabla q(x^{\ell+1}), d^\ell \rangle$ and the conjugacy of the vectors $d^{\ell+1}, \dots, d^k$ to d^ℓ,

$$\langle \nabla q(x^{k+1}), d^\ell \rangle = \langle \nabla q(x^{k+1}) - \nabla q(x^{\ell+1}), d^\ell \rangle = \sum_{j=\ell+1}^{k} t_e^j (d^j)^{\mathsf{T}} A d^\ell = 0.$$

$\square$

Theorem 2.2.59 *Let the vectors $d^0, \dots, d^{n-1}$ be pairwise conjugate with respect to A and all different from zero. Then x^n is the global minimal point of q.*

Proof According to Exercise 2.2.57, the vectors d^ℓ, $0 \le \ell \le n-1$, are linearly independent, and according to Lemma 2.2.58, it holds

$$\forall\, 0 \le \ell \le n-1: \quad \langle \nabla q(x^n), d^\ell \rangle = 0.$$

Therefore, the vector $\nabla q(x^n)$ vanishes, and x^n is the unique global minimal point of q (Exercise 2.1.43). $\square$

Theorem 2.2.59 states that a descent method for the convex-quadratic function q with exact step size control and pairwise conjugate search directions finds the global minimal point of q after at most n steps ('at most', because some x^k with $k < n$ may, by chance, already be a minimal point). In particular, since, due to $f(x^{k+1}) < f(x^k)$ for descent methods, $t_e^k \cdot d^k = x^{k+1} - x^k \ne 0$ applies, none of the vectors d^k vanish.

In the next step, we look for ways to explicitly generate conjugate search directions. The following theorem states that conjugate directions can be obtained, for example, from the quasi-Newton methods of the Broyden family.

Theorem 2.2.60 *For $\theta \ge 0$ apply Algorithm 2.6 with $t^k = t_e^k$ and $B^{k+1} = B_\theta^{k+1}$ to $q(x) = \frac{1}{2}x^{\mathsf{T}}Ax + b^{\mathsf{T}}x$ with $A = A^{\mathsf{T}} \succ 0$, and for some $k \in \mathbb{N}$ let the iterates $x^0, \dots, x^k$ be pairwise different. Then the directions $d^0, \dots, d^{k-1}$ are pairwise conjugate with respect to A and all different from zero.*

Proof We first establish the following relationships: By definition, for all $k \in \mathbb{N}$ we have

$$d^k = -B^k \nabla q(x^k), \tag{2.17}$$

$$x^{k+1} = x^k + t_e^k d^k \tag{2.18}$$

and thus

$$s^k := x^{k+1} - x^k = t_e^k d^k, \tag{2.19}$$

$$y^k := \nabla q(x^{k+1}) - \nabla q(x^k) = A s^k. \tag{2.20}$$

Furthermore, the secant equation (2.14) is equivalent to

$$B^{k+1} y^k = s^k, \tag{2.21}$$

and the exactness of the step size yields

$$\langle \nabla q(x^{k+1}), d^k \rangle = \varphi'_{d^k}(t_e^k) = 0. \tag{2.22}$$

The Broyden family methods with $\theta \geq 0$ are descent methods according to Lemma 2.2.55. As long as such a method does not terminate, it generates pairwise different iterates, so that the assumption of the theorem is fulfilled for the beginning of each sequence (x^k) generated by the method. Due to (2.18), all directions $d^0, \ldots, d^{k-1}$ are different from zero.

Next we show the following three-part auxiliary claim, the first part of which coincides with the main claim to be shown:

$$\forall 0 \leq \ell \leq k - 1: \qquad (d^k)^\mathsf{T} A d^\ell = 0, \tag{2.23}$$

$$\langle \nabla q(x^k), d^\ell \rangle = 0, \tag{2.24}$$

$$B^k y^\ell = s^\ell. \tag{2.25}$$

For $\ell = k - 1$ one sees this claim as follows: Due to $x^k \neq x^{k-1}$ and (2.18), t_e^{k-1} cannot disappear. Thus we have

$$(d^k)^\mathsf{T} A d^{k-1} \overset{(2.19)}{=} (d^k)^\mathsf{T} A \frac{s^{k-1}}{t_e^{k-1}} \overset{(2.17), (2.20)}{=} -\frac{1}{t_e^{k-1}} \langle \nabla q(x^k), B^k y^{k-1} \rangle$$

$$\overset{(2.21)}{=} -\frac{1}{t_e^{k-1}} \langle \nabla q(x^k), s^{k-1} \rangle \overset{(2.19)}{=} -\langle \nabla q(x^k), d^{k-1} \rangle \overset{(2.22)}{=} 0.$$

From this follows (2.23) and (2.24). Equation (2.25) coincides for $\ell = k - 1$ with the secant equation (2.21).

We handle the remaining cases by induction over k. For $k = 1$, the auxiliary assertion only needs to be shown for $\ell = 0$ which, in view of $\ell = k - 1$, has just been done. Let the auxiliary assertion now apply for some $k \in \mathbb{N}$, and let us prove it for $k + 1$. So we need to show the properties (2.23)–(2.25) with $k + 1$ instead of k for all $0 \leq \ell \leq k$. To distinguish the properties to be shown from those of the induction hypothesis, we will rather address them with $(2.23)_{k+1}$–$(2.25)_{k+1}$ and $(2.23)_k$–$(2.25)_k$, respectively.

The case $\ell = k$ has already been proven by the above consideration. So let $0 \leq \ell \leq k - 1$. It holds

$$(y^k)^\mathsf{T} B^k y^\ell \overset{(2.25)_k}{=} (y^k)^\mathsf{T} s^\ell \overset{(2.20)}{=} (s^k)^\mathsf{T} A s^\ell \overset{(2.19)}{=} t_e^k t_e^\ell (d^k)^\mathsf{T} A d^\ell \overset{(2.23)_k}{=} 0$$

and analogously

$$(s^k)^\mathsf{T} y^\ell = (s^k)^\mathsf{T} A s^\ell = 0.$$

From this follows

$$B^{k+1}y^\ell = B_\theta^{k+1}y^\ell = B^k y^\ell + \underbrace{\frac{s^k(s^k)^\mathsf{T}y^\ell}{(s^k)^\mathsf{T}y^k}}_{=\,0} - \underbrace{\frac{B^k y^k (y^k)^\mathsf{T} B^k y^\ell}{(y^k)^\mathsf{T} B^k y^k}}_{=\,0} + \theta r^k \underbrace{(r^k)^\mathsf{T}y^\ell}_{=\,0} \overset{(2.25)_k}{=} s^\ell$$

and thus $(2.25)_{k+1}$. Furthermore, we have

$$(d^{k+1})^\mathsf{T} A d^\ell = \frac{1}{t_e^\ell}(d^{k+1})^\mathsf{T} A s^\ell = -\frac{1}{t_e^\ell}\langle \nabla q(x^{k+1}), \underbrace{B^{k+1}y^\ell}_{=\,s^\ell}\rangle = -\langle \nabla q(x^{k+1}), d^\ell \rangle,$$

so that $(2.23)_{k+1}$ is also proven, as soon as we have shown $(2.24)_{k+1}$. Because of

$$\langle \nabla q(x^{k+1}), d^\ell \rangle = \langle A(x^k + t_e^k d^k) + b, d^\ell \rangle = \langle \nabla q(x^k), d^\ell \rangle + t_e^k (d^k)^\mathsf{T} A d^\ell$$

the latter follows from the induction assumptions $(2.23)_k$ and $(2.24)_k$.   □

When choosing exact step sizes, the quasi-Newton methods of the Broyden family thus minimize convex-quadratic functions in at most n steps. For an arbitrary C^2-function f, this can be interpreted as minimizing the local quadratic approximation to f in n steps, hence simulating one step of the Newton method in view of Exercise 2.2.46. Under suitable conditions and with restarts after every n steps, they therefore converge 'n-step-quadratically' (for details see [26]).

2.2.9 Conjugate Gradient Methods

As already mentioned, for large scale problems one actually wants to avoid storing matrices like B^k from the quasi-Newton methods. Therefore, one is looking for *matrix-free* ways to generate conjugate directions. Methods that iteratively generate conjugate directions are called *conjugate gradient methods* or briefly *CG methods*.

We continue to consider the convex-quadratic function

$$q(x) = \tfrac{1}{2}x^\mathsf{T} A x + b^\mathsf{T} x$$

with $A = A^\mathsf{T} \succ 0$ as well as $b \in \mathbb{R}^n$ and form the sequence (x^k) with exact step sizes. We are looking for ways to generate conjugate search directions (d^k).

Having constructed the iterates x^k recursively from previous iterates and, in the case of quasi-Newton methods, the matrices B^k recursively from previous matrices, the basic idea will be to also choose the search directions d^k recursively, namely as a combination of the current negative gradient $-\nabla q(x^k)$ and the previous search direction d^{k-1}. With the help of a yet to be determined 'weight' $\alpha_k \in \mathbb{R}$ this takes the form

$$d^k = -\nabla q(x^k) + \alpha_k \cdot d^{k-1}, \quad k = 1, 2, \dots$$

We initialize this recursion with

$$d^0 = -\nabla q(x^0).$$

The following lemma will be essential for determining appropriate values α_k.

Lemma 2.2.61 *Let $d^0, \ldots, d^{k-1}$ be pairwise conjugate with respect to A and $x^1, \ldots, x^k$ already generated with $x^\ell \neq x^{\ell-1}$ for $1 \leq \ell \leq k$. Then d^k is conjugate to some d^ℓ with $0 \leq \ell \leq k-1$ if and only if*

$$\langle \nabla q(x^{\ell+1}) - \nabla q(x^\ell), d^k \rangle = 0$$

is fulfilled.

Proof The assertion follows from

$$\langle \nabla q(x^{\ell+1}) - \nabla q(x^\ell), d^k \rangle = t_e^\ell (d^\ell)^\mathsf{T} A d^k$$

and $t_e^\ell \neq 0$.   □

Theorem 2.2.62 *Under the conditions of Lemma 2.2.61, the direction $d^k = -\nabla q(x^k) + \alpha_k \cdot d^{k-1}$ is conjugate to the vectors $d^0, \ldots, d^{k-1}$ if and only if*

$$\alpha_k = \frac{\|\nabla q(x^k)\|_2^2}{\|\nabla q(x^{k-1})\|_2^2}$$

is true.

Proof We prove the assertion by induction over k. In the case $k = 1$, according to Lemma 2.2.61 d^1 is conjugate to d^0 if and only if

$$0 = \langle \nabla q(x^1) - \nabla q(x^0), d^1 \rangle = \langle \nabla q(x^1) + d^0, -\nabla q(x^1) + \alpha_1 d^0 \rangle$$
$$= -\|\nabla q(x^1)\|_2^2 + \alpha_1 \|\nabla q(x^0)\|_2^2$$

holds, where the last equality follows from double application of (2.8).

For $k > 1$, we only show the conjugacy of d^k and d^{k-1} and leave the remaining cases as an exercise for the reader. According to Lemma 2.2.61, d^k is conjugate to d^{k-1} if and only if

$$0 = \langle \nabla q(x^k) - \nabla q(x^{k-1}), d^k \rangle = \langle \nabla q(x^k) - \nabla q(x^{k-1}), -\nabla q(x^k) + \alpha_k d^{k-1} \rangle$$

$$= -\|\nabla q(x^k)\|_2^2 + \underbrace{\langle \nabla q(x^{k-1}), \nabla q(x^k) \rangle}_{=: T_1} - \alpha_k \underbrace{\langle q(x^{k-1}), d^{k-1} \rangle}_{=: T_2}$$

holds. From

$$T_1 = \langle \nabla q(x^k), \alpha_{k-1} d^{k-2} - d^{k-1} \rangle$$

$$= \alpha_{k-1} \langle \nabla q(x^{k-1}) + t_e^{k-1} A d^{k-1}, d^{k-2} \rangle \overset{\text{ind. hyp.}}{=} 0$$

and

$$T_2 = \langle \nabla q(x^{k-1}), -\nabla q(x^{k-1}) + \alpha_{k-1} d^{k-2} \rangle = -\|\nabla q(x^{k-1})\|_2^2$$

the assertion follows. $\qquad\qquad\square$

Algorithm 2.7: Fletcher-Reeves method

Input: C^1-optimization problem P, starting point x^0 and termination tolerance $\varepsilon > 0$
Output: Approximation $\bar{x}$ of a critical point of f (if the method terminates [26, Th. 5.7])

1 **begin**
2   Set $d^0 = -\nabla f(x^0)$ and $k = 0$.
3   **while** $\|\nabla f(x^k)\| > \varepsilon$ **do**
4   Set $x^{k+1} = x^k + t_e^k d^k$.
5   Set $d^{k+1} = -\nabla f(x^{k+1}) + \left(\|\nabla f(x^{k+1})\|_2^2 / \|\nabla f(x^k)\|_2^2 \right) \cdot d^k$.
6   Replace k with $k + 1$.
7   **end**
8   Set $\bar{x} = x^k$.
9 **end**

Theorem 2.2.62 motivates Algorithm 2.7, as it provides the global minimal point for $f(x) = q(x) = \frac{1}{2} x^\mathsf{T} A x + b^\mathsf{T} x$ with $A = A^\mathsf{T} \succ 0$ after at most n steps. This method is used, for example, to solve large scale linear systems of equations $Ax = b$ using the least squares approach, i.e., by minimizing $\|r(x)\|_2^2$ with the residual $r(x) = Ax - b$.

Due to rounding errors (especially due to the division by $\|\nabla f(x^k)\|_2^2$), the method is rarely observed to terminate after n steps, so also its convergence speed was examined. It turns out that it depends on the square root of the condition number (i.e., the quotient of the largest and smallest eigenvalue) of the matrix $A^\mathsf{T} A$. It is therefore advisable to first transform the system of equations $Ax = b$ equivalently in such a way that this condition number decreases. This technique is known as *preconditioning*. For a brief introduction, see [16].

Remark 2.2.63 The least squares approach using the CG method to solve linear systems of equations $Ax = b$ can also be applied to *overdetermined* systems of equations that have no solution. One is then satisfied with some x that minimizes $\|r(x)\|_2^2$, even if the minimal value is not zero.

For *underdetermined* systems of equations, on the other hand, there is the possibility of selecting a special solution from all solutions of $Ax = b$. Often, the 'smallest' x is chosen in the sense that $\|x\|_2$ is minimized over the solution space. However, this leads to a constrained optimization problem (Chap. 3).

Crucial for the applicability of Algorithm 2.7 is that for $f = q$ the matrix A is not explicitly involved, but still search directions conjugate with respect to A are generated. The method can therefore also be formulated for arbitrary C^1-functions, where in line 4 the exact step size t_e^k generally has to be replaced by an inexact one like the Armijo step size t_a^k. Under suitable conditions, one again obtains that n CG steps simulate one Newton step, leading to 'n-step-quadratic convergence'. For this, it is recommended to do a 'restart' after every n steps by setting $d^{k \cdot n} = -\nabla f(x^{k \cdot n})$ for $k \in \mathbb{N}$.

Finally, in the case $f = q$, the relationships

$$
\frac{\|\nabla f(x^{k+1})\|_2^2}{\|\nabla f(x^k)\|_2^2} = \frac{\langle \nabla f(x^{k+1}), \nabla f(x^{k+1}) - \nabla f(x^k) \rangle}{\langle d^k, \nabla f(x^{k+1}) - \nabla f(x^k) \rangle}
$$

$$
= \frac{\langle \nabla f(x^{k+1}), \nabla f(x^{k+1}) - \nabla f(x^k) \rangle}{\|\nabla f(x^k)\|_2^2}
$$

are readily verified. For $f \neq q$, these equalities are not necessarily valid, so that replacing $\|\nabla f(x^{k+1})\|_2^2 / \|\nabla f(x^k)\|_2^2$ with these expressions in Algorithm 2.7 results in different methods, namely the CG methods of Hestenes-Stiefel and of Polak-Ribière, respectively (for details see [26]).

At the end of this section, we briefly discuss the *limited memory BFGS method* (*L-BFGS*) as an alternative to CG methods. It approximates BFGS updates (Sect. 2.2.7) using vectors of length n, without storing matrices. The matrices B^k already used as approximations of the inverse Hessian matrices $D^2 f(x^k)$ are thus approximated once more.

To illustrate the basic idea of this approach, we first note that the BFGS update from (2.16) can alternatively be written as

$$
B_{\mathrm{BFGS}}^{k+1} = \left(I - \frac{(s^k)(y^k)^\mathsf{T}}{(s^k)^\mathsf{T}(y^k)} \right) B^k \left(I - \frac{(s^k)(y^k)^\mathsf{T}}{(s^k)^\mathsf{T}(y^k)} \right) + \frac{(s^k)(s^k)^\mathsf{T}}{(s^k)^\mathsf{T}(y^k)}
$$

i.e., with a simple dependence on the current matrix B^k. If we replace the matrix B^k in this formula with its own definition as an update from the previous iteration, we obtain

$$
\begin{aligned}
B_{\text{BFGS}}^{k+1} &= \left(I - \frac{(s^k)(y^k)^\mathsf{T}}{(s^k)^\mathsf{T}(y^k)}\right)\left(I - \frac{(s^{k-1})(y^{k-1})^\mathsf{T}}{(s^{k-1})^\mathsf{T}(y^{k-1})}\right)B^{k-1} \\
&\quad \cdot \left(I - \frac{(s^{k-1})(y^{k-1})^\mathsf{T}}{(s^{k-1})^\mathsf{T}(y^{k-1})}\right)\left(I - \frac{(s^k)(y^k)^\mathsf{T}}{(s^k)^\mathsf{T}(y^k)}\right) \\
&\quad + \left(I - \frac{(s^k)(y^k)^\mathsf{T}}{(s^k)^\mathsf{T}(y^k)}\right)\frac{(s^{k-1})(s^{k-1})^\mathsf{T}}{(s^{k-1})^\mathsf{T}(y^{k-1})}\left(I - \frac{(s^k)(y^k)^\mathsf{T}}{(s^k)^\mathsf{T}(y^k)}\right) + \frac{(s^k)(s^k)^\mathsf{T}}{(s^k)^\mathsf{T}(y^k)}.
\end{aligned}
$$

The desired approximation of this update using vectors is achieved by replacing the usually dense (and therefore memory-intensive) matrix B^{k-1} with a sparse matrix like the identity matrix I or its scaling $((s^k)^\mathsf{T}(y^k)/(y^k)^\mathsf{T}(y^k))I$. The latter leads to the L-BFGS update of 'memory length two', thus to

$$
\begin{aligned}
B_{\text{L-BFGS},2}^{k+1} &= \frac{(s^k)^\mathsf{T}(y^k)}{(y^k)^\mathsf{T}(y^k)}\left(I - \frac{(s^k)(y^k)^\mathsf{T}}{(s^k)^\mathsf{T}(y^k)}\right)\left(I - \frac{(s^{k-1})(y^{k-1})^\mathsf{T}}{(s^{k-1})^\mathsf{T}(y^{k-1})}\right) \\
&\quad \cdot \left(I - \frac{(s^{k-1})(y^{k-1})^\mathsf{T}}{(s^{k-1})^\mathsf{T}(y^{k-1})}\right)\left(I - \frac{(s^k)(y^k)^\mathsf{T}}{(s^k)^\mathsf{T}(y^k)}\right) \\
&\quad + \left(I - \frac{(s^k)(y^k)^\mathsf{T}}{(s^k)^\mathsf{T}(y^k)}\right)\frac{(s^{k-1})(s^{k-1})^\mathsf{T}}{(s^{k-1})^\mathsf{T}(y^{k-1})}\left(I - \frac{(s^k)(y^k)^\mathsf{T}}{(s^k)^\mathsf{T}(y^k)}\right) + \frac{(s^k)(s^k)^\mathsf{T}}{(s^k)^\mathsf{T}(y^k)}.
\end{aligned}
$$

The recursive extension of this approach allows L-BFGS updates $B_{\text{L-BFGS},m}^{k+1}$ with arbitrary memory lengths m, where lengths of about $m = 10$ have proven effective.

Since the matrix $B_{\text{L-BFGS},m}^{k+1}$ itself is not required algorithmically, but only its product with the current gradient of the objective function, $d^{k+1} = -B_{\text{L-BFGS},m}^{k+1}\nabla f(x^{k+1})$, there is also a simple recursive procedure for determining the search direction. For details, see [26]. There, also the surprising connection is discussed that the 'memoryless' update $B_{\text{L-BFGS},1}^{k+1}$ in connection with exact step lengths leads to the CG methods of Hestenes-Stiefel and Polak-Ribière.

2.2.10 Trust-Region Methods

In contrast to classical line search methods, trust-region methods *first* choose the search radius t and *then* the search direction d (the formal dependence is therefore $d = d(t)$ instead of $t = t(d)$). To achieve this, in iteration k of the general descent method from Algorithm 2.3, a *quadratic model* for f is used as follows.

According to Taylor's theorem (Theorem 2.1.30b), for $f \in C^2(\mathbb{R}^n, \mathbb{R})$ we have the approximation property

$$
f(x^k + d) \approx f(x^k) + \langle \nabla f(x^k), d \rangle + \tfrac{1}{2}d^\mathsf{T}D^2 f(x^k)d.
$$

With $c^k := f(x^k)$, $b^k = \nabla f(x^k)$ and a symmetric matrix A^k (for example, but not necessarily, $A^k = D^2 f(x^k)$), the function

$$
m^k(d) := c^k + \langle b^k, d \rangle + \tfrac{1}{2}d^\mathsf{T}A^k d
$$

is called a *local quadratic model* for f around x^k. This terminology is motivated by the equalities $m^k(0) = f(x^k)$ and $\nabla m^k(0) = \nabla f(x^k)$ as well as the fact that m^k generally only describes the function f well for d from a sufficiently small neighborhood of the origin.

Therefore, m^k is only considered for $\|d\|_2 \leq t^k$ with a sufficiently small search radius t^k. If the behavior of m^k on $B_{\leq}(0, t^k)$ is 'good', then $B_{\leq}(0, t^k)$ is a 'trustworthy neighborhood' and actually called a *trust region*. To quantify the term 'good', a minimal point d^k of the *trust-region subproblem*

$$TR^k : \qquad \min_{d \in \mathbb{R}^n} m^k(d) \quad \text{s.t.} \quad \|d\|_2 \leq t^k$$

is determined. The quotient

$$r^k := \frac{f(x^k) - f(x^k + d^k)}{m^k(0) - m^k(d^k)}$$

of actual and expected descent in the objective function value then provides a measure of the quality of the local model. The difference $m^k(0) - m^k(d^k)$ is positive, even if the matrix A^k is not positive definite. In fact, due to the feasibility of $d = 0$ for TR^k the inequality $m^k(0) - m^k(d^k) \leq 0$ would imply the identity of $m^k(0)$ and $m^k(d^k)$, so that in addition to d^k also $d = 0$ would be a minimal point of TR^k. Since the constraint of TR^k is not active at $d = 0$, this would result in $0 = \nabla m^k(0) = \nabla f(x^k)$, contradicting the fact that the general descent method (Algorithm 2.3) would have already terminated in this case.

A value $r^k < 0$ therefore implies $f(x^k + d^k) > f(x^k)$, i.e., the update $x^{k+1} = x^k + d^k$ would result in an increase in the objective function value. Consequently, the trust region is too large, and its radius t^k must be reduced.

On the other hand, if r^k is close to one, then the local model describes the function f well; we set $x^{k+1} = x^k + d^k$ and tentatively increase the trust-region radius t^k in the next iteration.

Details on this procedure, especially on the acceptance of steps and the change of t^k for other values of r^k, are given in Algorithm 2.8. In particular, for $r^k \geq 1/4$, t^k is not reduced and the step is accepted, for $r^k < 0$, t^k is reduced and the step is rejected, and for $r^k \in [0, 1/4)$, t^k is reduced and the step is rejected if $r^k \leq \eta$ holds.

Figure 2.9 shows a possible new iterate x_{TR}^{k+1} of a trust-region method compared to a new iterate of a line search method x_{LS}^{k+1} with the same step size. At least in this example, the trust-region step brings us closer to the minimal point of f than the line search step.

Depending on the choice of matrices A^{k+1} in line 19 of Algorithm 2.8, one speaks of trust-region Newton methods, trust-region BFGS methods, etc. A crucial advantage of trust-region methods over variable metric line search methods is that the matrices A^k do not need to be positive definite.

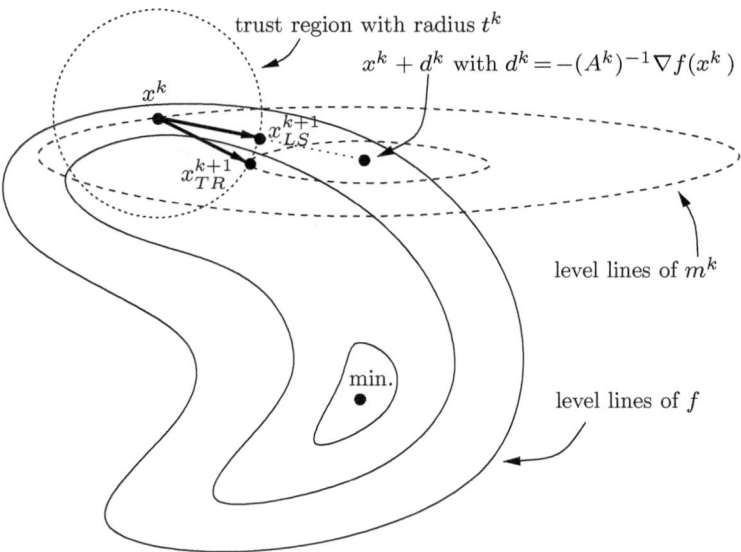

trust region with radius t^k

$x^k + d^k$ with $d^k = -(A^k)^{-1}\nabla f(x^k)$

x^k

x^{k+1}_{LS}

x^{k+1}_{TR}

level lines of m^k

min.

level lines of f

Fig. 2.9 Steps of line search and trust-region methods

In particular, for $A^k \equiv 0$ we would obtain a 'trust-region gradient method' with the trust-region subproblem

$$TR^k: \quad \min_{d \in \mathbb{R}^n} m^k(d) = f(x^k) + \langle \nabla f(x^k), d \rangle \quad \text{s.t.} \quad \|d\|_2 \le t^k.$$

Using, for example, the considerations from Sect. 2.1.3, its minimal point turns out to be

$$d^k = -\frac{t^k}{\|\nabla f(x^k)\|_2} \nabla f(x^k),$$

so that here only a usual gradient method with a special step size control arises. Due to Theorem 2.2.25, fast convergence is not to be expected from such a method, and trust-region gradient methods are thus not used.

Instead let us consider the case of a symmetric matrix $A^k \neq 0$. Solving the problem TR^k can be difficult, especially for an indefinite matrix A^k. However, analogous to the idea of inexact one-dimensional minimization in line search methods, an appropriate *inexact* solution of TR^k is also sufficient to ensure global convergence.

Algorithm 2.8: Trust-region method

Input: C^1-optimization problem P, starting point x^0, starting matrix $A^0 = (A^0)^\mathsf{T}$,
maximal radius $\check{t} > 0$, starting radius $t^0 \in (0, \check{t})$, parameter $\eta \in [0, 1/4)$
and termination tolerance $\varepsilon > 0$

Output: Approximation $\bar{x}$ of a critical point of f (if the method terminates; Theorem 2.2.67)

1 **begin**
2   Set $k = 0$.
3   **while** $\|\nabla f(x^k)\|_2 > \varepsilon$ **do**
4   Calculate an (inexact) minimal point d^k of TR^k and set

$$r^k = \frac{f(x^k) - f(x^k + d^k)}{m^k(0) - m^k(d^k)}.$$

5   **if** $r^k < \frac{1}{4}$ **then**
6   Set $t^{k+1} = \frac{1}{4}\|d^k\|_2$.
7   **else**
8   **if** $r^k > \frac{3}{4}$ **and** $\|d^k\|_2 = t^k$ **then**
9   Set $t^{k+1} = \min\{2t^k, \check{t}\}$.
10   **else**
11   Set $t^{k+1} = t^k$.
12   **end**
13   **end**
14   **if** $r^k > \eta$ **then**
15   Set $x^{k+1} = x^k + d^k$.
16   **else**
17   Set $x^{k+1} = x^k$.
18   **end**
19   Choose $A^{k+1} = (A^{k+1})^\mathsf{T}$.
20   Replace k with $k + 1$.
21   **end**
22   Set $\bar{x} = x^k$.
23 **end**

One possibility for this is to significantly reduce the feasible set of TR^k and, for example, only allow nonnegative multiples of the search direction found in the above 'trust-region gradient method':

$$TR_C^k: \quad \min_{d,s} m^k(d) \quad \text{s.t.} \quad \|d\|_2 \le t^k, \quad d = s \cdot \left(-\frac{t^k}{\|\nabla f(x^k)\|_2} \cdot \nabla f(x^k) \right)$$

$$s \ge 0.$$

This problem (where the index C stems from the fact that the gradient method is also referred to as the *Cauchy method*) is equivalent to

$$\min_{s \in \mathbb{R}} m^k \left(-\frac{s \cdot t^k}{\|\nabla f(x^k)\|_2} \cdot \nabla f(x^k) \right) \quad \text{s.t.} \quad 0 \le s \le 1,$$

thus to

$$\min_{s\in\mathbb{R}} \widetilde{m}^k(s) \quad \text{s.t.} \quad 0 \le s \le 1$$

with

$$\widetilde{m}^k(s) := m^k\left(-\frac{s \cdot t^k}{\|\nabla f(x^k)\|_2} \cdot \nabla f(x^k)\right)$$

$$= f(x^k) - s \cdot t^k \|\nabla f(x^k)\|_2 + s^2 \cdot \frac{(t^k)^2}{2\|\nabla f(x^k)\|_2^2} Df(x^k)A^k\nabla f(x^k).$$

To solve this one-dimensional optimization problem, two cases must be distinguished.

Case 1: $Df(x^k)A^k\nabla f(x^k) \le 0$. Then for all $s \in [0, 1]$

$$\frac{d}{ds} \widetilde{m}^k(s) = \underbrace{-t^k\|\nabla f(x^k)\|_2}_{<0} + \underbrace{s \frac{(t^k)^2}{\|\nabla f(x^k)\|_2^2} Df(x^k)A^k\nabla f(x^k)}_{\le 0}$$

holds, so that the function $\widetilde{m}^k$ strictly decreases on $[0, 1]$. Thus its minimal point is $s^\star = 1$.

Case 2: $Df(x^k)A^k\nabla f(x^k) > 0$. In this case, the function $\widetilde{m}^k$ is convex-quadratic, and the minimal point is either its apex

$$s^\star = \frac{\|\nabla f(x^k)\|_2^3}{t^k Df(x^k)A^k\nabla f(x^k)}$$

or $s^\star = 1$, depending on which value is smaller.

Overall, the point

$$d_C^k := -\frac{s^k t^k}{\|\nabla f(x^k)\|_2}\nabla f(x^k)$$

with

$$s^k := \begin{cases} 1, & \text{if } Df(x^k)A^k\nabla f(x^k) \le 0 \\ \min\left\{\frac{\|\nabla f(x^k)\|_2^3}{t^k Df(x^k)A^k\nabla f(x^k)}, 1\right\}, & \text{otherwise} \end{cases}$$

solves the problem TR_C^k.

Definition 2.2.64 (Cauchy Point) The point $x_C^{k+1} = x^k + d_C^k$ is called Cauchy point to x^k and t^k.

Exercise 2.2.65 Show that the vector $d^k = d_C^k$ satisfies the inequality

$$m^k(0) - m^k(d^k) \geq c \cdot \|\nabla f(x^k)\|_2 \cdot \min\left\{ t^k, \frac{\|\nabla f(x^k)\|_2}{\|A^k\|_2} \right\} \qquad (2.26)$$

with $c = 1/2$.

Remark 2.2.66 In view of the feasibility of d_C^k for TR^k, the *exact* solution d_e^k of TR^k satisfies the inequality $m^k(d_e^k) \leq m^k(d_C^k)$ and thus, according to Exercise 2.2.65, also (2.26) with $c = 1/2$.

We can now formulate a theorem on global convergence [26, Th. 4.5 and 4.6].

Theorem 2.2.67 *Let the set $f_{\leq}^{f(x^0)}$ be bounded, let the function ∇f be Lipschitz continuous on $\mathrm{conv}(f_{\leq}^{f(x^0)})$, let the sequence $(\|A^k\|_2)$ be bounded, and let the sequence (d^k) of inexact solutions of TR^k satisfy (2.26) with $c > 0$. Then the following holds in Algorithm 2.8:*

(a) *For $\eta = 0$, $\liminf_k \|\nabla f(x^k)\|_2 = 0$ is true (i.e., (x^k) has an accumulation point $x^\star$ with $\nabla f(x^\star) = 0$).*
(b) *For $\eta \in (0, 1/4)$, $\lim_k \nabla f(x^k) = 0$ is true (i.e., all accumulation points of (x^k) are critical).*

By Exercise 2.2.65, Remark 2.2.66, and Theorem 2.2.67, both the inexact solutions d_C^k and the exact solutions d_e^k of TR^k provide global convergence. While, as mentioned, the exact solution d_e^k may be difficult to calculate, resorting to the inexact solution d_C^k is rarely advisable, as the matrix A^k only influences the length of d_C^k and thus essentially still leads to the gradient method.

We therefore consider two commonly used inexact approximations for a minimal point d^k of TR^k, for which also $m^k(d^k) \leq m^k(d_C^k)$ and thus (2.26) with $c = 1/2$ hold (for details on convergence statements see [26]). The first approach possesses the positive definiteness of the matrices A^k as a restrictive assumption, while the second approach does not.

Dogleg Method

In the following, let $k \in \mathbb{N}$ be fixed and omitted and A be symmetric and positive definite. We start by asking how the exact solutions $d_e(t)$ of *TR* behave with varying $t \geq 0$. By definition, $d_e(t)$ solves the problem

$$TR: \quad \min_{d \in \mathbb{R}^n} c + b^\mathsf{T} d + \tfrac{1}{2} d^\mathsf{T} A d \quad \text{s.t.} \quad \|d\|_2 \leq t$$

with $c = f(x)$ and $b = \nabla f(x)$. For t close to zero, the term $d^\mathsf{T} A d$ is negligible compared to $b^\mathsf{T} d$, so that $d_e(t)$ then approximately coincides with the solution of

$$\min_{d \in \mathbb{R}^n} c + b^\mathsf{T} d \quad \text{s.t.} \quad \|d\|_2 \leq t,$$

thus with $-t \nabla f(x)/\|\nabla f(x)\|_2$. If t is sufficiently large on the other hand, the *unconstrained* minimal point of $c + b^\mathsf{T} d + \tfrac{1}{2} d^\mathsf{T} A d$, namely

$$d_A := -A^{-1} \nabla f(x),$$

fulfills the constraint $\|d\|_2 \leq t$ of *TR*. Hence, for increasing $t \geq 0$ the points $x + d_e(t)$ can be expected to describe a curve from x to $x + d_A$ as shown in Fig. 2.10.

The dogleg method approximates this curve by a polygonal line from x to $x + d_A$ with two segments, a 'dogleg path'. The intermediate point $x + d_G$ is chosen to be the exact minimal point d_G of m along $-\nabla f(x)$, which is calculated as

$$d_G = -\frac{\|\nabla f(x)\|_2^2}{Df(x) A \nabla f(x)} \nabla f(x).$$

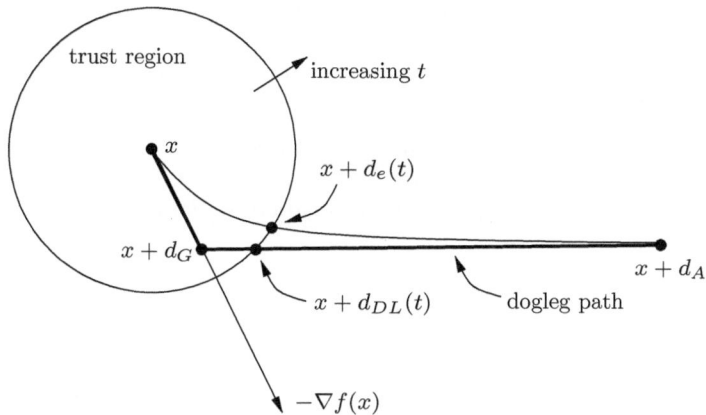

Fig. 2.10 Approximation of the curve $\{x + d_e(t) \,|\, t \geq 0\}$ by a dogleg path

Formally, the dogleg path is thus parametrized as $\{\, x + \widetilde{d}(s) \mid s \in [0, 2]\,\}$ with

$$\widetilde{d}(s) \;=\; \begin{cases} s \cdot d_G\,, & 0 \le s \le 1 \\ d_G + (s-1)(d_A - d_G)\,, & 1 \le s \le 2. \end{cases}$$

Exercise 2.2.68 Show for $A = A^{\mathsf{T}} \succ 0$:

(a) $\|\widetilde{d}(s)\|_2$ is monotonically increasing in s.
(b) $m(\widetilde{d}(s))$ is monotonically decreasing in s.

By Exercise 2.2.68a, in the case $t > \|d_A\|_2$ the dogleg path does not intersect the boundary of the trust region at all, and otherwise at exactly one point. This is a consequence of

$$\|s \cdot d_G\|_2 \;=\; t\,, \qquad\qquad \text{if } t < \|d_G\|_2$$
$$\|d_G + (s-1)(d_A - d_G)\|_2 \;=\; t\,, \text{ otherwise.}$$

From Exercise 2.2.68b it follows that the solution of the problem *TR*, constrained to the dogleg path, is just d_A in the case $t > \|d_A\|_2$, while in the case $t \le \|d_A\|_2$ it is the above intersection of the dogleg path with the boundary of the trust region. The respective solution is the inexact solution $d_{DL}(t)$ chosen by the dogleg method for *TR*, with which the new iterate $x + d_{DL}(t)$ is generated.

Minimization on a Two-Dimensional Subspace
The inexact solution of *TR* by the dogleg method can be improved by not restricting *TR* to the *one*-dimensional dogleg path, but to the *two*-dimensional subspace, which is spanned by d_G and d_A. This space, in particular, contains the dogleg path, but is a less coarse substitute for the feasible set of *TR*. One obtains the auxiliary problem

$$\min_{d \in \mathbb{R}^n} m(d) \quad \text{s.t.} \quad \|d\|_2 \le t\,, \quad d \in \mathrm{img}(\nabla f(x),\, A^{-1}\nabla f(x)),$$

thus a problem in two variables, which is again explicitly solvable.

A main advantage of this approach is that, in contrast to the dogleg method, it can be meaningfully extended to indefinite matrices A. For details, see [26].

Constrained Optimization

3

Contents

Already the examples in Sect. 1.1, the calculation of spectral norms (Remark 2.2.41), the determination of norm-minimal solutions of linear systems of equations (Remark 2.2.63) as well as the trust-region subproblem (Sect. 2.2.10) show that, when optimizing an objective function, constraints often need to be considered. We distinguish between inequality and equality constraints and consider the general form of constrained optimization problems

$$P: \quad \min_{x \in \mathbb{R}^n} f(x) \quad \text{s.t.} \quad g_i(x) \le 0, \ i \in I, \quad h_j(x) = 0, \ j \in J,$$

O. Stein, *Basic Concepts of Nonlinear Optimization*,
Mathematics Study Resources 8, https://doi.org/10.1007/978-3-662-69741-2_3

with at least continuous functions f, g_i, $h_j : \mathbb{R}^n \to \mathbb{R}$ for $i \in I$ and $j \in J$. The index sets I and J are assumed to be finite and possibly empty, and we set

$$I = \{1, \ldots, p\} \quad \text{and} \quad J = \{1, \ldots, q\}$$

with $p, q \in \mathbb{N} \cup \{0\}$. Furthermore, we assume $q < n$, because under the linear independence constraint qualification, which is typically fulfilled for equality constraints (Definition 3.2.43), q equations in $\mathbb{R}^n$ define an $(n - q)$-dimensional manifold. Thus, for $q = n$ or $q > n$, the equality constraints would usually define a discrete or empty feasible set, while the theory and algorithms of continuous optimization assume a feasible set that is at least one-dimensional. On the other hand, an upper bound on the number p of inequality constraints does not need to be assumed. In semi-infinite optimization (Example 1.1.5), even infinitely many inequalities are allowed.

Although for simplicity the entire space $\mathbb{R}^n$ is assumed as the domain of the objective function f in problem P, in many of the following considerations it could be restricted to the feasible set M. Differentiability requirements such as $f \in C^1(M, \mathbb{R})$ then require, however, that f is defined (and continuously differentiable) on some open superset of M.

Section 3.1 provides some properties of the feasible set needed for deriving optimality conditions, before Sect. 3.2 states various first and second-order optimality conditions based on these. Afterwards, Sect. 3.3 discusses important solution methods for constrained nonlinear optimization problems.

3.1 Properties of the Feasible Set

The definition of the vector-valued functions

$$g(x) := \begin{pmatrix} g_1(x) \\ \vdots \\ g_p(x) \end{pmatrix} \quad \text{and} \quad h(x) := \begin{pmatrix} h_1(x) \\ \vdots \\ h_q(x) \end{pmatrix}$$

allows the set M of *feasible points* for P to be written briefly as

$$M = \{x \in \mathbb{R}^n \mid g(x) \le 0, \ h(x) = 0\},$$

where the inequality between the p-dimensional vectors $g(x)$ and 0 is to be understood componentwise. We will occasionally make use of this description of M in the following.

To understand the optimality conditions discussed in Sect. 3.2, it will be important to know basic properties of the feasible set M. These include the *topological* properties discussed in Sect. 3.1.1, which are linked to the *continuity* of the functions g and h defining the set M, and the possibilities for defining *first-order approximations* to M discussed in Sect. 3.1.2, which are related, among other things, to first-order differentiability properties of g and h. These approximations will play a central role in Sect. 3.2 for formulating a suitable stationarity condition.

3.1.1 Topological Properties

We first note some important topological properties of the set M and conclusions from them. In Exercise 1.2.11, we have already seen that M is a *closed* set under the continuity assumption for the functions g and h.

In the following, we examine more closely the *activity* of inequality constraints, as already mentioned in Example 1.1.1. Here, an inequality constraint $g_i(x) \leq 0$ is called *active* at a feasible point $\bar{x}$ if it is satisfied there with equality, i.e., with $g_i(\bar{x}) = 0$. If the strict inequality $g_i(\bar{x}) < 0$ holds, the inequality constraint $g_i(x) \leq 0$ is instead called *inactive* at $\bar{x}$. *Equality* constraints are active at all feasible points in this sense.

> **Definition 3.1.1 (Active Index Set)** For $\bar{x} \in M$, the set
>
> $$I_0(\bar{x}) = \{i \in I \mid g_i(\bar{x}) = 0\}$$
>
> is called the *set of active indices* or briefly *active index set*.

A more precise (but also more cumbersome) name for $I_0(\bar{x})$ would be 'set of indices of the inequality constraints active at $\bar{x}$'.

Example 3.1.2 In the problem from Example 1.1.1, let (e.g., for marketing reasons) the additional constraint $r \geq 1$ be introduced. Then in the optimal radius $\bar{r}$, the new inequality is

- active for $\sqrt{A/(6\pi)} \leq 1 \leq \sqrt{A/(2\pi)}$, because $\bar{r} = 1$ holds, and
- inactive for $1 < \sqrt{A/(6\pi)}$, because $\bar{r} = \sqrt{A/(6\pi)} > 1$ holds.

Figure 3.1 illustrates that, for a local description of the set M around a point $\bar{x} \in M$, the active constraints should suffice, that is, in the general case the constraints $g_i(x) \leq 0$, $i \in I_0(\bar{x})$, and $h_j(x) = 0$, $j \in J$. Indeed, the following result holds.

Fig. 3.1 Local description of M through active constraints

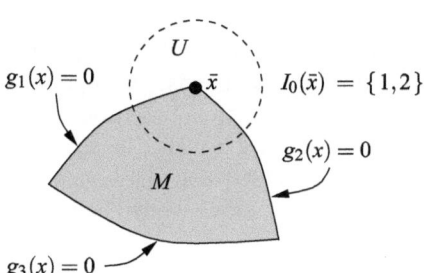

Theorem 3.1.3 *For each $\bar{x} \in M$ there exists some neighborhood U of $\bar{x}$ with*

$$U \cap M = U \cap \{x \in \mathbb{R}^n | \ g_i(x) \leq 0, \ i \in I_0(\bar{x}), \ h_j(x) = 0, \ j \in J\}.$$

Proof The inclusion $\subseteq$ is clear for every neighborhood U of $\bar{x}$. To see the reverse inclusion $\supseteq$, we assume that there is no neighborhood U of $\bar{x}$ with

$$U \cap M \supseteq U \cap \{x \in \mathbb{R}^n | \ g_i(x) \leq 0, \ i \in I_0(\bar{x}), \ h_j(x) = 0, \ j \in J\}.$$

Then for every neighborhood U of $\bar{x}$ there exists some $x_U \in U$ with

$$g_i(x_U) \leq 0, \ i \in I_0(\bar{x}), \ h_j(x_U) = 0, \ j \in J,$$

but $x_U \notin M$. In particular, for every $k \in \mathbb{N}$ and for the neighborhood $U_k := B_{\leq}(\bar{x}, 1/k)$ there exists some $x^k \in U_k$ with

$$g_i(x^k) \leq 0, \ i \in I_0(\bar{x}), \ h_j(x^k) = 0, \ j \in J$$

and $x^k \notin M$. From the specific choice of neighborhoods $\lim_k x^k = \bar{x}$ follows.

For a given $k \in \mathbb{N}$, $x^k \notin M$ can only hold if there exists some $i_k \in I \setminus I_0(\bar{x})$ with $g_{i_k}(x^k) > 0$. For taking the limit $k \to \infty$ in these expressions, the dependence of the index i on k constitutes an obstacle. We free i from this dependence by the following consideration.

Suppose each index in $I \setminus I_0(\bar{x})$ is only 'hit' finitely often by the sequence (i_k). Since $I \setminus I_0(\bar{x})$ contains only finitely many elements, the sequence (i_k) and thus also (x^k) would have to stop after finitely many terms. But since this is not the case, at least one index $i_0 \in I \setminus I_0(\bar{x})$ appears infinitely often in the sequence (i_k). We switch to the subsequence formed by the x^k with $i_k = i_0$ (again without subsequence notation) and thus obtain

$$\exists \, i_0 \in I \setminus I_0(\bar{x}) \quad \forall k \in \mathbb{N}: \quad g_{i_0}(x^k) > 0.$$

From the continuity of g_{i_0}, the inequality $g_{i_0}(\bar{x}) \geq 0$ follows in the limit $k \to \infty$. On the other hand, $\bar{x} \in M$ also implies $g_{i_0}(\bar{x}) \leq 0$, so overall $g_{i_0}(\bar{x}) = 0$ and thus $i_0 \in I_0(\bar{x})$ are true. However, this contradicts the choice $i_0 \in I \setminus I_0(\bar{x})$. $\qquad \square$

In Sect. 3.2, we will develop a first-order optimality condition for constrained optimization problems, analogous to the derivation of Fermat's rule (Theorem 2.1.13), which is based on the fact that no descent direction for the objective function f can exist at a local minimal point. In the constrained case, however, it must be added that only such descent directions can be excluded at a local minimal point, along which the feasible set M is not left. This leads to the following definition.

Definition 3.1.4 (Feasible Descent Direction) For the problem

$$P: \quad \min f(x) \quad \text{s.t.} \quad x \in M$$

with feasible set $M \subseteq \mathbb{R}^n$ (not necessarily given in functional description), a vector $d \in \mathbb{R}^n$ is called a *feasible descent direction* for P at $\bar{x} \in M$, if

$$\exists \check{t} > 0 \quad \forall t \in (0, \check{t}): \quad f(\bar{x} + td) < f(\bar{x}), \quad \bar{x} + td \in M$$

holds.

Exercise 3.1.5 For the problem P from Definition 3.1.4, let $\bar{x}$ be a local minimal point. Show that then no feasible descent direction for P exists at $\bar{x}$.

If we additionally assume that M is described by inequality and equality constraints as

$$M = \{x \in \mathbb{R}^n \mid g_i(x) \le 0, \ i \in I, \ h_j(x) = 0, \ j \in J\},$$

then with the help of Theorem 3.1.3 the next statement can be proven, according to which for the property of a vector d to be a feasible descent direction for P at $\bar{x}$ only the *active* constraints at $\bar{x}$ need to be checked.

Exercise 3.1.6 Consider the problem

$$P: \quad \min f(x) \quad \text{s.t.} \quad g_i(x) \le 0, \ i \in I, \ h_j(x) = 0, \ j \in J.$$

Show that a vector $d \in \mathbb{R}^n$ is a feasible descent direction for P at $\bar{x} \in M$ if and only if

$$\exists \check{t} > 0 \quad \forall t \in (0, \check{t}):$$

$$f(\bar{x} + td) < f(\bar{x}), \quad g_i(\bar{x} + td) \le 0, \ i \in I_0(\bar{x}), \quad h_j(\bar{x} + td) = 0, \ j \in J,$$

holds.

3.1.2   First-Order Approximations

To derive a first-order necessary optimality condition, as in the unconstrained case we will exploit in Sect. 3.2 that in a local minimal point a stationarity condition applies, i.e., the existence of certain feasible descent directions is excluded, which arise from *first-order* information. Since the linearization of the objective function f has already been extensively discussed in Sect. 2.1, the present section focuses on first-order approximations of the feasible set M.

The explicit treatment of equality constraints would be just technical in the following, so temporarily we only consider purely inequality-constrained problems (i.e., with $J = \emptyset$). In the following we assume the functions f and g_i, $i \in I$, to be at least differentiable at the point $\bar{x} \in \mathbb{R}^n$ under consideration.

Gradients of the involved functions naturally occur through their linearizations around a given point $\bar{x}$. For example, Theorem 2.1.30a provides, by omitting the error term, the first-order approximation $f(\bar{x}) + \langle \nabla f(\bar{x}), x - \bar{x} \rangle$ of f locally around $\bar{x}$ as well as for each $i \in I$ the linearization $g_i(\bar{x}) + \langle \nabla g_i(\bar{x}), x - \bar{x} \rangle$ of g_i around $\bar{x}$. It is therefore natural to consider the following linearization of the constrained optimization problem P around a point $\bar{x}$:

$$P_{\mathrm{lin}}(\bar{x}): \quad \min_{x \in \mathbb{R}^n} \; f(\bar{x}) + \langle \nabla f(\bar{x}), x - \bar{x} \rangle \quad \text{s.t.} \quad g_i(\bar{x}) + \langle \nabla g_i(\bar{x}), x - \bar{x} \rangle \le 0,$$

$$i \in I_0(\bar{x}),$$

where due to Theorem 3.1.3 instead of the complete index set I only $I_0(\bar{x})$ is used. Because of $g_i(\bar{x}) = 0$, $i \in I_0(\bar{x})$, and since according to Exercise 1.3.1 the constant $f(\bar{x})$ is irrelevant for the determination of minimal points of $P_{\mathrm{lin}}(\bar{x})$, the substitution $d := x - \bar{x}$ simplifies this problem to

$$P_{\mathrm{lin}}(\bar{x}): \quad \min_{d \in \mathbb{R}^n} \; \langle \nabla f(\bar{x}), d \rangle \quad \text{s.t.} \quad \langle \nabla g_i(\bar{x}), d \rangle \le 0, \; i \in I_0(\bar{x}).$$

This motivates the following definition.

> **Definition 3.1.7 (Outer Linearization Cone)** For $\bar{x} \in M$, the set
>
> $$L_\le(\bar{x}, M) = \{d \in \mathbb{R}^n \,|\, \langle \nabla g_i(\bar{x}), d \rangle \le 0, \; i \in I_0(\bar{x})\}$$
>
> is called the *outer linearization cone* to M at $\bar{x}$.

The terminology is explained by the fact that we will also introduce an *inner* linearization cone in Definition 3.1.11. A set $A \subseteq \mathbb{R}^n$ is called a *cone* if

$$\forall\, a \in A, \, \lambda > 0: \quad \lambda \cdot a \in A$$

applies. The fact that sets of direction vectors possess this property is not surprising, because scaling with positive numbers does not change the direction into which a vector points.

Exercise 3.1.8 At the point $\bar{x} \in M$, let the functions g_i, $i \in I_0(\bar{x})$, be differentiable. Show that then $L_\le(\bar{x}, M)$ is a convex cone.

Fig. 3.2 Two outer
linearization cones in
Example 3.1.9

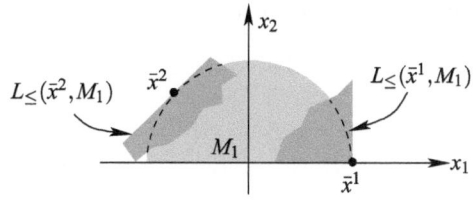

Example 3.1.9 For $n=p=2$, the set M_1 defined by the inequalities

$$g_1(x) = x_1^2 + x_2^2 - 1 \leq 0,$$

$$g_2(x) = -x_2 \leq 0$$

is shown in Fig. 3.2. At the point $\bar{x}^1 = (1,0)^\mathsf{T}$, both inequalities are active. Evaluating their gradients $\nabla g_1(x) = 2x$ and $\nabla g_2(x) = (0,-1)^\mathsf{T}$ at $\bar{x}^1$ yields $\nabla g_1(\bar{x}^1) = (2,0)^\mathsf{T}$ and $\nabla g_2(\bar{x}^1) = (0,-1)^\mathsf{T}$, respectively, so that the outer linearization cone is

$$L_\leq(\bar{x}^1, M_1) = \{d \in \mathbb{R}^2 | (2,0)\, d \leq 0, \ (0,-1)\, d \leq 0\}$$

$$= \{d \in \mathbb{R}^2 | d_1 \leq 0, \ d_2 \geq 0\}.$$

In Fig. 3.2, due to the general relation $d = x - \bar{x}$, for better geometric visualization the set $\bar{x}^1 + L_\leq(\bar{x}^1, M_1)$ is shown, but for simplicity it is labeled as $L_\leq(\bar{x}^1, M_1)$.

At the point $\bar{x}^2 = 1/\sqrt{2}(-1,1)^\mathsf{T}$, only the constraint g_1 is active, and one obtains

$$L_\leq(\bar{x}^2, M_1) = \{d \in \mathbb{R}^2 | \sqrt{2}(-1,1)\, d \leq 0\} = \{d \in \mathbb{R}^2 | d_2 \leq d_1\}.$$

In Example 3.1.9, the outer linearization cones at both $\bar{x}^1$ and $\bar{x}^2$ apparently reflect the local structure of the set well 'to first order'. In particular, with an objective function like $f(x) = x_2 - x_1$, both the nonlinear problem P and its linearization $P_{\mathrm{lin}}(\bar{x}^1)$ possess a minimal point at $\bar{x}^1$. Unfortunately, this is *not* always the case, as the next example shows.

Example 3.1.10 In Example 3.1.9, let the function $g_2(x) = -x_2$ be replaced by $\widetilde{g}_2(x) = -x_2^3$. It is easy to see that the set

$$\widetilde{M}_1 = \{x \in \mathbb{R}^2 | g_1(x) \leq 0, \ \widetilde{g}_2(x) \leq 0\}$$

geometrically coincides with the set M_1 from Example 3.1.9. However, this does *not* apply to the outer linearization cone $L_\leq(\bar{x}^1, \widetilde{M}_1)$: Due to $\nabla \widetilde{g}_2(x) = (0, -3x_2^2)^\mathsf{T}$ and $\nabla \widetilde{g}_2(\bar{x}^1) = (0,0)^\mathsf{T}$, one obtains

$$L_\leq(\bar{x}^1, \widetilde{M}_1) = \{d \in \mathbb{R}^2 | (2,0)\, d \leq 0, \ (0,0)\, d \leq 0\} = \{d \in \mathbb{R}^2 | d_1 \leq 0\}$$

(Fig. 3.3).

Fig. 3.3 Outer linearization
cone in Example 3.1.10

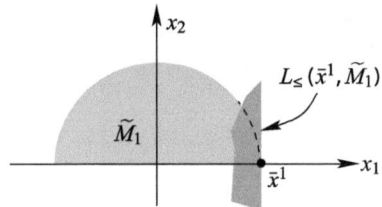

Example 3.1.10 shows that the *functional description* of a feasible set can be
so unsuitable that an outer linearization cone does not necessarily reflect the local
structure of the set well. In particular, the problem $P_{\text{lin}}(\bar{x})$ does not provide a good
local description of the problem P around $\bar{x}$. Particularly disturbing for our aim to
derive first-order optimality conditions is that an objective function like $f(x) = x_2 -
x_1$ in Example 3.1.10 does possess the minimal point $\bar{x}^1$ in the nonlinear problem $\widetilde{P}$
(which is defined using $\widetilde{g}_2$ instead of g_2), but the point $\bar{x}^1$ is *not* a minimal point of
the linearization $\widetilde{P}_{\text{lin}}(\bar{x}^1)$.

Whether the functional description is 'good' or 'bad' can be determined by
comparing the outer with an inner linearization cone, which is defined as follows.

> **Definition 3.1.11 (Inner Linearization Cone)** For $\bar{x} \in M$, the set
>
> $$L_<(\bar{x}, M) = \{d \in \mathbb{R}^n | \langle \nabla g_i(\bar{x}), d \rangle < 0, \ i \in I_0(\bar{x})\}$$
>
> is called the *inner linearization cone* to M at $\bar{x}$.

While $L_\leq(\bar{x}, M)$ is a closed convex cone that contains the origin, $L_<(\bar{x}, M)$
is an open convex cone not containing the origin. Although the two cones are
different from each other, they seem to differ only 'marginally'. To specify this
more precisely, let clA denote the (topological) *closure* of a set A (i.e., the set of limit
points of all convergent sequences $(x^k) \subseteq A$). Because of $L_<(\bar{x}, M) \subseteq L_\leq(\bar{x}, M)$
and the closedness of $L_\leq(\bar{x}, M)$, the inclusion cl$L_<(\bar{x}, M) \subseteq L_\leq(\bar{x}, M)$ is clear.

> **Definition 3.1.12 (Nondegenerate Functional Description of a Set)** The
> functional description of M is called *nondegenerate* at $\bar{x} \in M$ if
> cl$L_<(\bar{x}, M) = L_\leq(\bar{x}, M)$ is true. Otherwise, it is called *degenerate*.

The equality cl$L_<(\bar{x}, M) = L_\leq(\bar{x}, M)$ is known as the *Cottle constraint
qualification*.

Example 3.1.13 In Example 3.1.9 it holds

$$L_<(\bar{x}^1, M_1) = \{d \in \mathbb{R}^2 | d_1 < 0, \ d_2 > 0\},$$
$$L_<(\bar{x}^2, M_1) = \{d \in \mathbb{R}^2 | d_2 < d_1\}$$

and thus

$$\text{cl}L_<(\bar{x}^1, M_1) = \{d \in \mathbb{R}^2 | d_1 \leq 0, \ d_2 \geq 0\} = L_\leq(\bar{x}^1, M_1)$$

as well as

$$\text{cl}L_<(\bar{x}^2, M_1) = \{d \in \mathbb{R}^2 | d_2 \leq d_1\} = L_\leq(\bar{x}^2, M_1).$$

The functional description of M_1 is thus nondegenerate at both $\bar{x}^1$ and $\bar{x}^2$.

Example 3.1.14 In Example 3.1.10 it holds

$$L_<(\bar{x}^1, \widetilde{M}_1) = \{d \in \mathbb{R}^2 | (2,0)\,d < 0, \ (0,0)\,d < 0\} = \emptyset,$$
$$L_<(\bar{x}^2, \widetilde{M}_1) = \{d \in \mathbb{R}^2 | d_2 < d_1\}$$

and thus

$$\text{cl}L_<(\bar{x}^1, \widetilde{M}_1) = \emptyset \subsetneq \{d \in \mathbb{R}^2 | d_1 \leq 0\} = L_\leq(\bar{x}^1, M_1)$$

as well as

$$\text{cl}L_<(\bar{x}^2, \widetilde{M}_1) = \{d \in \mathbb{R}^2 | d_2 \leq d_1\} = L_\leq(\bar{x}^2, \widetilde{M}_1).$$

The functional description of $\widetilde{M}_1$ is thus degenerate at $\bar{x}^1$ and nondegenerate at $\bar{x}^2$.

The fact that the inner linearization cone is empty in the case of a degenerate functional description as in Example 3.1.14 is no coincidence, as the next theorem shows. It allows to replace the topological and therefore algorithmically difficult to verify Cottle constraint qualification $\text{cl}L_<(\bar{x}, M) = L_\leq(\bar{x}, M)$ by a purely algebraic and therefore algorithmically easier to check condition.

Theorem 3.1.15 *The functional description of M is nondegenerate at $\bar{x} \in M$ if and only if $L_<(\bar{x}, M) \neq \emptyset$ holds.*

Proof First, let $L_<(\bar{x}, M) = \emptyset$. This implies $\mathrm{cl}L_<(\bar{x}, M) = \emptyset$, while $L_\leq(\bar{x}, M)$ contains at least the point $d = 0$. The functional description of M at $\bar{x}$ is thus degenerate.

On the other hand, let $L_<(\bar{x}, M) \neq \emptyset$. We have to show the equality $\mathrm{cl}L_<(\bar{x}, M) = L_\leq(\bar{x}, M)$, of which the inclusion $\mathrm{cl}L_<(\bar{x}, M) \subseteq L_\leq(\bar{x}, M)$ always applies, as already mentioned. To prove the reverse inclusion, let $\bar{d} \in L_\leq(\bar{x}, M)$. We choose some $\varepsilon > 0$ and $d^0 \in L_<(\bar{x}, M)$, whose existence is guaranteed by the assumption. For each $i \in I_0(\bar{x})$, it then holds

$$\langle \nabla g_i(\bar{x}), \bar{d} + \varepsilon d^0 \rangle = \underbrace{\langle \nabla g_i(\bar{x}), \bar{d} \rangle}_{\leq 0} + \varepsilon \underbrace{\langle \nabla g_i(\bar{x}), d^0 \rangle}_{< 0} < 0.$$

Consequently, $\bar{d} + \varepsilon d^0$ lies in $L_<(\bar{x}, M)$. Taking the limit $\varepsilon \to 0$ yields $\bar{d} \in \mathrm{cl}L_<(\bar{x}, M)$. $\qquad\square$

There are also cases where the *geometry* of the feasible set is so awkward that *no* functional description provides the desired 'good' first-order approximation.

Example 3.1.16 For $n = p = 2$, the set M_2 defined by the inequality constraints

$$g_1(x) = (x_1 - 1)^3 + x_2 \leq 0,$$
$$g_2(x) = -x_2 \leq 0$$

is shown in Fig. 3.4. At the point $\bar{x}^1 = (1, 0)^\mathsf{T}$ both inequalities are active. Evaluating their gradients $\nabla g_1(x) = (3(x_1 - 1)^2, 1)^\mathsf{T}$ and $\nabla g_2(x) = (0, -1)^\mathsf{T}$ at $\bar{x}^1$ yields $\nabla g_1(\bar{x}^1) = (0, 1)^\mathsf{T}$ and $\nabla g_2(\bar{x}^1) = (0, -1)^\mathsf{T}$, which yields the outer linearization cone

$$L_\leq(\bar{x}^1, M_2) = \{d \in \mathbb{R}^2 | (0, 1)d \leq 0, (0, -1)d \leq 0\} = \{d \in \mathbb{R}^2 | d_2 = 0\}$$

and the inner linearization cone $L_<(\bar{x}^1, M_2) = \emptyset$. Thus, the functional description of M_2 at $\bar{x}^1$ is degenerate.

In Corollary 3.1.26 we will see that in Example 3.1.16 *every* functional description of M_2 at $\bar{x}^1$ must be degenerate. For this, we introduce alternative local

Fig. 3.4 Outer linearization cone in Example 3.1.16

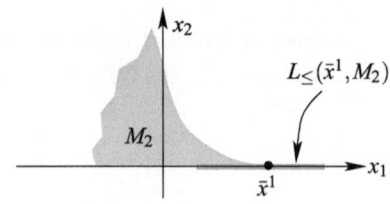

first-order approximations of M that *only* depend on the geometry of the set M and *not* on its functional description.

Definition 3.1.17 (Inner and Outer Tangent Cones) For a not necessarily closed set $X \subseteq \mathbb{R}^n$ let $\bar{x} \in \mathrm{cl}X$. A direction $\bar{d} \in \mathbb{R}^n$ lies in the

(a) *inner tangent cone* $\Gamma(\bar{x}, X)$ to X at $\bar{x}$, if some $\check{t} > 0$ and some neighborhood D of $\bar{d}$ exist with

$$\forall\, t \in (0, \check{t}),\ d \in D : \bar{x} + td \in X,$$

(b) *outer tangent cone* $C(\bar{x}, X)$ to X at $\bar{x}$, if sequences (t^k) and (d^k) exist with

$$t^k \searrow 0,\ d^k \to \bar{d},\ \forall\, k \in \mathbb{N} : \bar{x} + t^k d^k \in X.$$

Figure 3.5 shows examples of inner and outer tangent cones to a closed set M as well as vectors $\bar{d}^1 \notin \Gamma(\bar{x}^1, M)$ and $\bar{d}^2 \in C(\bar{x}^2, M)$. If we did not allow variable direction vectors in Definition 3.1.17, this would result in $\bar{d}^1 \in \Gamma(\bar{x}^1, M)$ and $\bar{d}^2 \notin C(\bar{x}^2, M)$. We want to avoid this so that the inner and outer tangent cones to a closed set M possess the following similar properties as the corresponding linearization cones. Note that in Definition 3.1.17 we allow X to be nonclosed so that the statement of part b in the following result makes sense.

Lemma 3.1.18 *For a closed set $M \subseteq \mathbb{R}^n$ and $\bar{x} \in M$ the following assertions are true.*

(a) $\Gamma(\bar{x}, M) \subseteq C(\bar{x}, M)$.
(b) $\Gamma(\bar{x}, M)^c = C(\bar{x}, M^c)$.
(c) $\Gamma(\bar{x}, M)$ *is an open and* $C(\bar{x}, M)$ *a closed cone.*

Fig. 3.5 Inner and outer tangent cone

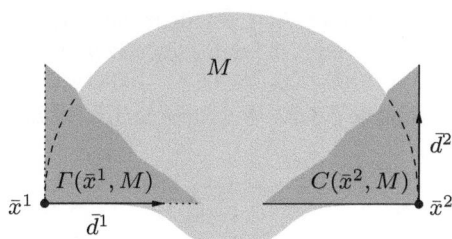

Proof Let $\bar{d}$ be an arbitrary element of the set $\Gamma(\bar{x}, M)$ with corresponding $\check{t} > 0$ and a neighborhood D of $\bar{d}$. Then, with some $k_0 \in \mathbb{N}$, for all $k \geq k_0$

$$t^k := \frac{1}{k} \in (0, \check{t}), \qquad d^k := \bar{d} \in D$$

and thus $\bar{x} + t^k d^k \in M$ hold. From $t^k \searrow 0$ and $d^k \to \bar{d}$ follows $\bar{d} \in C(\bar{x}, M)$, thus statement a.

To see the inclusion $\subseteq$ in statement b, we use that a vector $\bar{d}$ does not lie in $\Gamma(\bar{x}, M)$ if for all $\check{t} > 0$ and all neighborhoods D of $\bar{d}$ some $t \in (0, \check{t})$ and $d \in D$ with $\bar{x} + td \in M^c$ exist. In particular, there then exist for $\overset{*}{t} = 1/k$ and $D_k = B_{\leq}(\bar{d}, 1/k)$ some $t^k \in (0, \overset{*}{t})$ and $d^k \in D_k$ with $\bar{x} + t^k d^k \in M^c$. Because of $t^k \searrow 0$ and $d^k \to \bar{d}$ it follows $\bar{d} \in C(\bar{x}, M^c)$. The proof of the inclusion $\supseteq$ in statement b is left as an exercise to the reader.

To show in statement c that $\Gamma(\bar{x}, M)$ is open, we choose $\bar{d} \in \Gamma(\bar{x}, M)$ with corresponding $\check{t} > 0$ and an open neighborhood D of $\bar{d}$. For all $\tilde{d} \in D$ there exists some neighborhood $\tilde{D}$ of $\tilde{d}$ with $\tilde{D} \subseteq D$. Thus, for all $t \in (0, \check{t})$ and $d \in \tilde{D}$ the point $\bar{x} + td$ lies in M. This means $\tilde{d} \in \Gamma(\bar{x}, M)$ and finally $D \subseteq \Gamma(\bar{x}, M)$.

The closedness of $C(\bar{x}, M)$ follows from the openness of $\Gamma(\bar{x}, M)$ and statement b, and the cone property of $\Gamma(\bar{x}, M)$ as well as $C(\bar{x}, M)$ is left to the reader as an exercise. □

Because of $\Gamma(\bar{x}, M) \subseteq C(\bar{x}, M)$ and the closedness of $C(\bar{x}, M)$, the inclusion $\mathrm{cl}\,\Gamma(\bar{x}, M) \subseteq C(\bar{x}, M)$ is clear, in analogy to the respective result for the linearization cones.

> **Definition 3.1.19 (Nondegenerate Geometry of a Set)** The geometry of M is called *nondegenerate* at $\bar{x}$ if $\mathrm{cl}\,\Gamma(\bar{x}, M) = C(\bar{x}, M)$ is true. Otherwise, it is called *degenerate*.

Example 3.1.20 In Example 3.1.9, the identities $\Gamma(\bar{x}^1, M_1) = L_<(\bar{x}^1, M_1)$, $C(\bar{x}^1, M_1) = L_\leq(\bar{x}^1, M_1)$, $\Gamma(\bar{x}^2, M_1) = L_<(\bar{x}^2, M_1)$ and $C(\bar{x}^2, M_1) = L_\leq(\bar{x}^2, M_1)$ hold. From the considerations in Example 3.1.13 it follows that the geometry of M_1 is nondegenerate at both $\bar{x}^1$ and $\bar{x}^2$.

Example 3.1.21 In Example 3.1.10, although the functional description of $\widetilde{M}_1$ differs from that of M_1, $\widetilde{M}_1 = M_1$ holds, so the geometry of both sets is identical. Since the tangent cones only depend on the geometry, Example 3.1.20 provides the nondegeneracy of the geometry of $\widetilde{M}_1$ at both $\bar{x}^1$ and $\bar{x}^2$.

According to Example 3.1.14, however, the *functional description* of $\widetilde{M}_1$ at $\bar{x}^1$ is degenerate. At $\bar{x}^1$, neither the inner linearization cone matches the inner tangent cone, nor does the outer linearization cone match the outer tangent cone.

Example 3.1.22 In Example 3.1.16, $L_<(\bar{x}^1, M_2) = \Gamma(\bar{x}^1, M_2) = \emptyset$ and

$$C(\bar{x}^1, M_2) = \{d \in \mathbb{R}^2 | \, d_1 \leq 0, \, d_2 = 0\} \neq \{d \in \mathbb{R}^2 | \, d_2 = 0\} = L_\leq(\bar{x}^1, M_2)$$

hold. In particular, the geometry of M_2 at $\bar{x}^1$ is degenerate.

A degenerate geometry does not always, as in Example 3.1.22, result in an *empty* inner tangent cone (Exercise 3.1.23), so the analogue to Theorem 3.1.15 for degenerate functional descriptions does *not* apply. However, for $\bar{x} \in M$ the outer tangent cone $C(\bar{x}, M)$ contains the point $d = 0$, so that the geometric degeneracy at least follows from $\Gamma(\bar{x}, M) = \emptyset$.

Exercise 3.1.23 Let $M = \{x \in \mathbb{R}^2 \mid x_1 x_2^2 \leq 0\}$ and $\bar{x} = 0$. Show

$$\emptyset \neq \{d \in \mathbb{R}^2 \mid d_1 \leq 0\} = \mathrm{cl}\Gamma(\bar{x}, M) \neq C(\bar{x}, M)$$

$$= \{d \in \mathbb{R}^2 \mid d_1 \leq 0\} \cup \{d \in \mathbb{R}^2 \mid d_2 = 0\}.$$

The set M is therefore geometrically degenerate at $\bar{x} = 0$ without $\Gamma(\bar{x}, M)$ being void. This example also shows that the outer tangent cone, unlike the outer linearization cone, is *not necessarily convex*.

To establish a connection between the degeneracy of the geometry and that of the functional description, we next examine the relationships between the four approximating cones more closely.

Theorem 3.1.24 *For all $\bar{x} \in M$ the chain of inclusions*

$$L_<(\bar{x}, M) \subseteq \Gamma(\bar{x}, M) \subseteq C(\bar{x}, M) \subseteq L_\leq(\bar{x}, M)$$

is true.

Proof Thanks to Lemma 3.1.18b, to prove the first inclusion it is sufficient to alternatively show $C(\bar{x}, M^c) \subseteq (L_<(\bar{x}, M))^c$. Let $\bar{d} \in C(\bar{x}, M^c)$, i.e., there exist sequences $t^k \searrow 0$ and $d^k \to \bar{d}$ with $\bar{x} + t^k d^k \notin M$ for all $k \in \mathbb{N}$.

For all $k \in \mathbb{N}$ there is therefore some $i_k \in I$ with $g_{i_k}(\bar{x} + t^k d^k) > 0$. Due to the finiteness of I, at least one index i_0 occurs infinitely often in the sequence (i_k). After passing to the corresponding subsequence, we obtain $g_{i_0}(\bar{x} + t^k d^k) > 0$ for all $k \in \mathbb{N}$ and in the limit $g_{i_0}(\bar{x}) \geq 0$. In view of $\bar{x} \in M$ we also have $g_{i_0}(\bar{x}) \leq 0$, so in total $i_0 \in I_0(\bar{x})$.

This implies by Taylor's theorem (Theorem 2.1.30a) for all $k \in \mathbb{N}$

$$0 < \frac{g_{i_0}(\bar{x} + t^k d^k) - g_{i_0}(\bar{x})}{t^k} = \frac{\langle \nabla g_{i_0}(\bar{x}), t^k d^k \rangle + o(\|t^k d^k\|)}{t^k}$$

$$= \langle \nabla g_{i_0}(\bar{x}), d^k \rangle + \omega(\bar{x} + t^k d^k)\|d^k\|,$$

which, due to the properties of the function ω, implies

$$0 \leq \langle \nabla g_{i_0}(\bar{x}), \bar{d} \rangle + \omega(\bar{x}) \|\bar{d}\| = \langle \nabla g_{i_0}(\bar{x}), \bar{d} \rangle$$

in the limit $k \to \infty$. Therefore, $\bar{d}$ cannot lie in $L_<(\bar{x}, M)$, which was to be shown.

The second claimed inclusion was already shown in Lemma 3.1.18a, and the proof of the third inclusion is left as an exercise for the reader.   □

Remark 3.1.25 The expression

$$\lim_k \frac{g_{i_0}(\bar{x} + t^k d^k) - g_{i_0}(\bar{x})}{t^k}$$

appearing in the proof of Theorem 3.1.24 is reminiscent of the one-sided directional derivative of g_{i_0} at $\bar{x}$ in the direction $\bar{d}$ from Definition 2.1.4,

$$g'_{i_0}(\bar{x}, \bar{d}) = \lim_{t \searrow 0} \frac{g_{i_0}(\bar{x} + t\bar{d}) - g_{i_0}(\bar{x})}{t}.$$

There, the direction $\bar{d} \in \mathbb{R}^n$ is *held constant* during the limiting process, while in the above proof we have performed a limit of the form

$$\lim_{t \searrow 0, \, d \to \bar{d}} \frac{g_{i_0}(\bar{x} + td) - g_{i_0}(\bar{x})}{t}.$$

If such an expression exists and is identical for all choices $t^k \searrow 0$ and $d^k \to \bar{d}$, it is referred to as a (one-sided) *directional derivative in the sense of Hadamard*, while the directional derivative from Definition 2.1.4 is called (one-sided) *directional derivative in the sense of Dini*.

In the proof of Theorem 3.1.24 we have thus shown, among other things, that differentiable functions are not only one-sided directionally differentiable in the sense of Dini, but also in the sense of Hadamard.

Corollary 3.1.26 *If the functional description of the set M is nondegenerate at $\bar{x} \in M$, then the geometry of M at $\bar{x}$ is also nondegenerate.*

Proof Due to Theorem 3.1.24 and the closedness of $C(\bar{x}, M)$, the nondegeneracy of the functional description of M at $\bar{x}$ implies

$$L_\leq(\bar{x}, M) \subseteq \text{cl}L_<(\bar{x}, M) \subseteq \text{cl}\Gamma(\bar{x}, M) \subseteq C(\bar{x}, M) \subseteq L_\leq(\bar{x}, M).$$

All inclusions of this chain must therefore be identities, and in particular $\text{cl}\Gamma(\bar{x}, M) = C(\bar{x}, M)$ holds, which was to be shown.   □

Example 3.1.27 Since in Example 3.1.16 the set M_2 is geometrically degenerate at the point $\bar{x}^1$, according to Corollary 3.1.26 M_2 cannot possess any nondegenerate functional description at $\bar{x}^1$.

The same applies to the set M from Exercise 3.1.23 at the point $\bar{x} = 0$.

3.2 Optimality Conditions

In this section, we derive the generalizations of first and second-order optimality conditions from the unconstrained to the constrained case. As a generalization of the central first-order condition $\nabla f(x) = 0$ from the unconstrained case, we will essentially obtain the Karush-Kuhn-Tucker conditions. The limitation 'essentially' refers to the fact that the feasible set M must meet certain regularity conditions, namely constraint qualifications. To make their meaning transparent and understandable, we present a detailed derivation of the Karush-Kuhn-Tucker conditions.

For this, Sect. 3.2.1 first provides a generalization of the concept of stationarity from the unconstrained to the constrained case. Since this concept depends only on the geometry of the feasible set (and not on its functional description), it is presented in an abstract form. To substantiate it with the help of the functional description of the feasible set, the constraint qualifications discussed in Sect. 3.2.2 are required.

This concretization of the stationarity condition results in the unsolvability of certain systems of inequalities. To be able to exploit the latter algorithmically, we characterize this unsolvability in Sect. 3.2.3 with the help of so-called theorems of the alternative. The separation theorem required for their proof is provided separately in Sect. 3.2.5 to avoid interrupting the line of arguments. Based on the theorems of the alternative, Sect. 3.2.4 provides the Karush-Kuhn-Tucker and Fritz-John conditions as first-order necessary optimality conditions. After proving the separation theorem in Sect. 3.2.5, Sect. 3.2.6 discusses an important general relationship between stationarity conditions and the concept of the normal cone.

Since equality constraints have been ignored in our derivation up to this point, Sect. 3.2.7 completes the derivation of optimality conditions by their presence. Section 3.2.8 discusses some important interpretations of the Karush-Kuhn-Tucker conditions and shows along an example that they can also be fulfilled at some points that are not locally minimal, analogous to the critical point condition in the unconstrained case. This motivates the consideration of second-order optimality conditions in Sect. 3.2.9. Finally, Sect. 3.2.10 shows (again analogous to the unconstrained case) that the Karush-Kuhn-Tucker conditions are also *sufficient* for optimality in *convex* optimization problems.

3.2.1 Stationarity

In the unconstrained case we have seen that, at a local minimal point, no descent direction in the sense of Definition 2.1.1 can exist (Exercise 2.1.2), so in particular also no first-order descent direction in the sense of Definition 2.1.7 exists (Lemma 2.1.6). This necessary optimality condition, i.e. the absence of a first-order descent direction, was called stationarity in Definition 2.1.8.

In this section we will proceed analogously for the constrained case. Since local minimality of a point depends only on the *geometry* of the feasible set M, we initially do not use any functional description of M. We already know from Exercise 3.1.5

that at a local minimal point of

$$P: \quad \min f(x) \quad \text{s.t.} \quad x \in M$$

there cannot exist *any* feasible descent direction in the sense of Definition 3.1.4. In order to employ this for 'first-order feasible descent directions', it is not enough to consider only a linear approximation of the objective function, but we also approximate the feasible set to first order using the tools from Sect. 3.1.2. Since we only want to use the geometry of M, we define stationarity using a *geometric* first-order approximation, namely the outer tangent cone.

Definition 3.2.1 (Stationary Point—Constrained Case) Let the function $f : \mathbb{R}^n \to \mathbb{R}$ be differentiable at $\bar{x} \in M$. Then $\bar{x}$ is called a *stationary point* of P, if $\langle \nabla f(\bar{x}), d \rangle \geq 0$ holds for every direction $d \in C(\bar{x}, M)$.

In the constrained case, stationarity at $\bar{x}$ thus means (somewhat cumbersome formulated) the absence of 'first-order geometrically feasible first-order descent directions'. The following result is the analogue to Lemma 2.1.6 from the unconstrained case.

Theorem 3.2.2 *Let the function* $f : \mathbb{R}^n \to \mathbb{R}$ *be differentiable at a local minimal point* $\bar{x}$ *of* P. *Then* $\bar{x}$ *is a stationary point in the sense of Definition 3.2.1.*

Proof Let $\bar{d} \in C(\bar{x}, M)$, so there are sequences $t^k \searrow 0$ and $d^k \to \bar{d}$ with $\bar{x} + t^k d^k \in M$ for all $k \in \mathbb{N}$. Since $\bar{x}$ is a local minimal point, with some $k_0 \in \mathbb{N}$ for all $k \geq k_0$ the inequality $f(\bar{x} + t^k d^k) \geq f(\bar{x})$ holds. From the positivity of t^k it also follows

$$\frac{f(\bar{x} + t^k d^k) - f(\bar{x})}{t^k} \geq 0$$

and

$$\langle \nabla f(\bar{x}), \bar{d} \rangle = \lim_k \frac{f(\bar{x} + t^k d^k) - f(\bar{x})}{t^k} \geq 0,$$

where we have again exploited the one-sided directional differentiability of the function f at $\bar{x}$ in the sense of Hadamard (Remark 3.1.25). □

3.2.2 Constraint Qualifications

Since the stationarity concept from Definition 3.2.1 involves the abstractly described outer tangent cone, the necessary optimality condition from Theorem 3.2.2 is not algorithmically accessible. To substantiate it, we return to a functional description of M and hope for a functional description of $C(\bar{x}, M)$. At first glance, the outer linearization cone $L_{\leq}(\bar{x}, M)$ seems suitable. However, Theorem 3.1.24 only provides that $L_{\leq}(\bar{x}, M)$ is a *super*set of $C(\bar{x}, M)$, and the Examples 3.1.21 and 3.1.22 (as well as Exercise 3.1.23) illustrate that the identity of both cones cannot be expected. Then one must not replace the stationarity condition with the condition $\langle \nabla f(\bar{x}), d \rangle \geq 0$ for all $d \in L_{\leq}(\bar{x}, M)$, because at local minimal points there may exist directions $d \in L_{\leq}(\bar{x}, M) \setminus C(\bar{x}, M)$ with $\langle \nabla f(\bar{x}), d \rangle < 0$ (as in Example 3.1.10 at $\bar{x}^1$ for $f(x) = x_2 - x_1$).

A possible remedy is to resort to the functional description of a *sub*set of $C(\bar{x}, M)$. For this, according to Theorem 3.1.24, the *inner* linearization cone $L_{<}(\bar{x}, M)$ is suitable. At a stationary point and thus also at every local minimal point of P, it is therefore necessarily true that $\langle \nabla f(\bar{x}), d \rangle \geq 0$ holds for all $d \in L_{<}(\bar{x}, M)$.

However, the Examples 3.1.14 and 3.1.16 (as well as Exercise 3.1.23) show that the inner linearization cone can be empty, so that the just formulated necessary optimality condition is then *trivially* fulfilled and thus does not provide a 'proper condition'.

These observations lead to the definition of two regularity conditions, called *constraint qualifications*.

Definition 3.2.3 (Abadie and Mangasarian-Fromovitz Constraint Qualifications for $J = \emptyset$) At $\bar{x} \in M$ holds

(a) the *Abadie constraint qualification (ACQ)* for $J = \emptyset$, if

$$C(\bar{x}, M) \ = \ L_{\leq}(\bar{x}, M)$$

   is fulfilled,

(b) the *Mangasarian-Fromovitz constraint qualification (MFCQ)* for $J = \emptyset$, if

$$L_{<}(\bar{x}, M) \ \neq \ \emptyset$$

   is true.

According to the definition of the inner linearization cone, the MFCQ at a point $\bar{x} \in M$ is fulfilled if and only if a direction $d \in \mathbb{R}^n$ with

$$\langle \nabla g_i(\bar{x}), d \rangle < 0, \quad i \in I_0(\bar{x}),$$

exists. Example 3.1.14 shows that the MFCQ can be violated at a point simply because for the geometrically nondegenerate feasible set a degenerate functional description is used. According to Theorem 3.1.15 and Corollary 3.1.26, however, the MFCQ is certainly violated at every point where the feasible set is geometrically degenerate.

Theorem 3.1.15 with the just introduced terminology provides a characterization of the nondegeneracy of the functional description by the validity of the MFCQ (or, in other words, the equivalence of the topologically formulated Cottle condition with the algebraically formulated MFCQ).

We summarize which conclusions can be drawn from the stationarity condition (identified as a necessary optimality condition in Theorem 3.2.2), when we replace the abstractly described outer tangent cone with one of the two functionally described linearization cones.

Corollary 3.2.4 *At a local minimal point $\bar{x}$ of P, let f and the functions g_i, $i \in I_0(\bar{x})$, be differentiable.*

(a) *Then the system*

$$\langle \nabla f(\bar{x}), d \rangle < 0, \quad \langle \nabla g_i(\bar{x}), d \rangle < 0, \quad i \in I_0(\bar{x}), \tag{3.1}$$

is unsolvable for any $d \in \mathbb{R}^n$.

(b) *If the ACQ holds at $\bar{x}$, then even the system*

$$\langle \nabla f(\bar{x}), d \rangle < 0, \quad \langle \nabla g_i(\bar{x}), d \rangle \leq 0, \quad i \in I_0(\bar{x}), \tag{3.2}$$

is unsolvable for any $d \in \mathbb{R}^n$.

Proof The proof follows from Theorems 3.1.24 and 3.2.2 as well as the definitions of the inner and outer linearization cones.   □

Exercise 3.2.5 Show that for $\bar{x} \in M$ every direction d with (3.1) is a feasible descent direction in $\bar{x}$ in the sense of Definition 3.1.4, both for P and for the linearized problem $P_{\text{lin}}(\bar{x})$.

Exercise 3.2.6 Show that for $\bar{x} \in M$ every direction d with (3.2) is a feasible descent direction for $P_{\text{lin}}(\bar{x})$ in $\bar{x}$ in the sense of Definition 3.1.4.

Exercise 3.2.7 Provide a problem P and a point $\bar{x} \in M$ where the ACQ holds, but there exists a direction d with (3.2) which is not a feasible descent direction for P in $\bar{x}$ in the sense of Definition 3.1.4.

Exercise 3.2.7 shows that stationarity even excludes the existence of certain *infeasible* descent directions.

Since the necessary condition from Corollary 3.2.4a can be trivially fulfilled when the MFCQ is violated, in the following we will first try to exploit the condition from Corollary 3.2.4b algorithmically. For its validity, however, one needs to check the ACQ, which is formulated by means of the abstractly described outer tangent cone. Hence, we need to know algorithmically verifiable sufficient conditions for the validity of the ACQ at a point $\bar{x} \in M$.

The following theorem shows that, for example, the MFCQ is such a sufficient condition for the ACQ.

Theorem 3.2.8 *At every $\bar{x} \in M$, the MFCQ implies the ACQ.*

Proof Let the MFCQ hold at some $\bar{x} \in M$. According to Theorem 3.1.15, M is then nondegenerately functionally described at $\bar{x}$. The rest of the proof proceeds verbatim as that of Corollary 3.1.26, except for the exploitation of the set identity $C(\bar{x}, M) = L_{\leq}(\bar{x}, M)$ in the final step. $\square$

Exercise 3.2.9 Let the function $f : \mathbb{R}^n \to \mathbb{R}$ be differentiable at $\bar{x}$ with $\nabla f(\bar{x}) \neq 0$, and consider the lower level set

$$f_{\leq}^{f(\bar{x})} = \{x \in \mathbb{R}^n \mid f(x) \leq f(\bar{x})\}$$

of f at the level $f(\bar{x})$. Prove the statement

$$C\left(\bar{x}, f_{\leq}^{f(\bar{x})}\right) = \{d \in \mathbb{R}^n \mid \langle \nabla f(\bar{x}), d \rangle \leq 0\}$$

claimed in Sect. 2.1.3.

In addition to Theorem 3.2.8, another important sufficient condition for the validity of the ACQ is motivated by the fact that the ACQ states a 'linearizability property' of the feasible set, which should hold automatically if the set is already described by finitely many linear inequalities. In the latter case M is called *polyhedral*.

Example 3.2.10 For all $1 \leq i \leq p$, let

$$g_i(x) = a_i^\mathsf{T} x + b_i$$

with $a_i \in \mathbb{R}^n$, $b_i \in \mathbb{R}$. In matrix-vector notation, $g(x) \le 0$ is then equivalent to the constraint $Ax + b \le 0$ known from linear optimization, where we define

$$A := \begin{pmatrix} a_1^\mathsf{T} \\ \vdots \\ a_p^\mathsf{T} \end{pmatrix} \quad \text{and} \quad b := \begin{pmatrix} b_1 \\ \vdots \\ b_p \end{pmatrix}.$$

We show that in this case the ACQ is fulfilled everywhere in M.

Indeed, let $\bar{x} \in M$ be arbitrary. Due to Theorem 3.1.24, we only need to show $L_\le(\bar{x}, M) \subseteq C(\bar{x}, M)$. For this, let $\bar{d} \in L_\le(\bar{x}, M)$, i.e., for all $i \in I_0(\bar{x})$ it holds

$$0 \ge \langle \nabla g_i(\bar{x}), \bar{d} \rangle = a_i^\mathsf{T} \bar{d}.$$

We need to show the existence of sequences $t^k \searrow 0$ and $d^k \to \bar{d}$, such that $\bar{x} + t^k d^k \in M$ is fulfilled for all $k \in \mathbb{N}$. We achieve this with the simple choices $t^k = 1/k$ and $d^k \equiv \bar{d}$, because it holds

$$\forall\, i \in I_0(\bar{x}),\ k \in \mathbb{N}: \quad g_i(\bar{x} + t^k d^k) = a_i^\mathsf{T}(\bar{x} + t^k d^k) + b_i$$
$$= \underbrace{a_i^\mathsf{T}\bar{x} + b_i}_{= 0} + \tfrac{1}{k}\underbrace{a_i^\mathsf{T}\bar{d}}_{\le 0} \le 0$$

and

$$\forall\, i \in I \setminus I_0(\bar{x}),\ k \ge k_i: \quad g_i(\bar{x} + t^k d^k) < 0$$

with some $k_i \in \mathbb{N}$, which exist in view of the continuity of g_i and $g_i(\bar{x}) < 0$. With $k_0 := \max_{i \in I \setminus I_0(\bar{x})} k_i$ we finally obtain

$$\forall\, i \in I,\ k \ge k_0: \quad g_i(\bar{x} + t^k d^k) \le 0$$

and thus $\bar{x} + t^k d^k \in M$ for all $k \ge k_0$.

So if $\bar{x} \in M$ is a local minimal point of a differentiable function f over such a polyhedral set M, then according to Corollary 3.2.4b the system

$$\langle \nabla f(\bar{x}), d \rangle < 0, \quad \langle a_i, d \rangle \le 0, \quad i \in I_0(\bar{x}),$$

cannot be solved with any $d \in \mathbb{R}^n$.

Exercise 3.2.11 Construct a feasible set $M \subseteq \mathbb{R}^2$, described by two inequalities, that satisfies the MFCQ nowhere, but the ACQ everywhere.

Exercise 3.2.12 In the description of the set $M = \{x \in \mathbb{R}^n \mid g_i(x) \leq 0, \, i \in I\}$ let all functions g_i, $i \in I$, be concave on $\mathbb{R}^n$. Show that then the ACQ holds at every point of M.

Another sufficient condition for the ACQ, which is even sufficient for the MFCQ, will be provided in Definition 3.2.43.

3.2.3 Theorems of the Alternative

To make the necessary optimality conditions from Corollary 3.2.4 algorithmically accessible, we use theorems that characterize the unsolvability of inequality systems, namely so-called *theorems of the alternative*.

For this, we return to considering the unsolvability of the system of strict inequalities (3.1) from Corollary 3.2.4a. Temporarily disregarding that the defining vectors of these inequalities are gradients of functions from an optimization problem, the condition in Corollary 3.2.4a states that with a number $r \in \mathbb{N}$ and certain vectors $a^1, \ldots, a^r$ the system $\langle a^k, d \rangle < 0$, $1 \leq k \leq r$, has no solution $d \in \mathbb{R}^n$. Geometrically, this means that there is no vector d that simultaneously forms an obtuse angle with all vectors $a^1, \ldots, a^r$.

On the right of Fig. 3.6, three vectors are drawn for which this is the case, while on the left a vector d exists that forms an obtuse angle with all three vectors at the same time. The figure also depicts the *convex hull* of these three vectors in both cases. In general, the convex hull conv(A) of a set $A \subseteq \mathbb{R}^n$ consists of the set of all *convex combinations* of elements in A, i.e.

$$\text{conv}(A) = \left\{ \sum_{i=1}^{s} \lambda_i a^i \, \middle| \, a^i \in A, \, \lambda_i \geq 0, \, 1 \leq i \leq s, \, \sum_{i=1}^{s} \lambda_i = 1, \, s \in \mathbb{N} \right\}.$$

In Fig. 3.6 one observes that the unsolvability of the system of inequalities (right) goes hand in hand with the origin being contained in the convex hull of the three vectors, while this is not the case when the system of inequalities is solvable (left). That this is indeed *always* true, ultimately leads to algorithmically exploitable optimality conditions and is the content of the following *central result*. Its proof requires some preparation.

Fig. 3.6 Obtuse angles and convex hulls

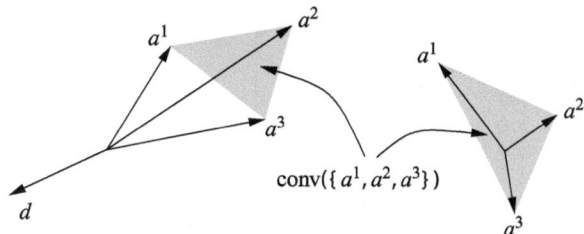

Theorem 3.2.13 (Gordan's Lemma) *For vectors $a^k \in \mathbb{R}^n$, $1 \leq k \leq r$, with $r \in \mathbb{N}$, exactly one of the following alternatives applies.*

(a) The system $\langle a^k, d \rangle < 0$, $1 \leq k \leq r$, possesses a solution $d \in \mathbb{R}^n$.
(b) It holds $0 \in \mathrm{conv}(\{a^1, \dots, a^r\})$.

Proof (Part 1) We first only show that from statement b the negation of statement a follows. Let $0 \in \mathrm{conv}(\{a^1, \dots, a^r\})$, so there are $\lambda_1, \dots, \lambda_r \geq 0$ with $\sum_{k=1}^r \lambda_k = 1$ and $0 = \sum_{k=1}^r \lambda_k a^k$. Assuming that also statement a is true, there exists some $d \in \mathbb{R}^n$ with $\langle a^k, d \rangle < 0$, $1 \leq k \leq r$. Then it follows

$$0 = \langle 0, d \rangle = \sum_{k=1}^r \lambda_k \underbrace{\langle a^k, d \rangle}_{< 0}.$$

Due to $\lambda_1, \dots, \lambda_r \geq 0$ this is only possible for $\lambda_k = 0$, $1 \leq k \leq r$, which however contradicts the requirement $\sum_{k=1}^r \lambda_k = 1$. Thus, statement a is false. □

We note that the formulation of Gordan's lemma using two mutually exclusive alternatives a and b has historical reasons. An equivalent formulation, tailored to the application of interest to us, states that the system $\langle a^k, d \rangle < 0$, $1 \leq k \leq r$, is *unsolvable* if and only if $0 \in \mathrm{conv}(\{a^1, \dots, a^r\})$ holds. The above first part of the proof shows the 'if direction' of this formulation, but this is unfortunately useless for the derivation of the necessary optimality condition we are interested in.

The second part of the proof of Theorem 3.2.13 is more profound, because it requires the application of the following *separation theorem*. We postpone its proof to Sect. 3.2.5, to first complete the proof of Theorem 3.2.13, and to be able to draw further conclusions.

Theorem 3.2.14 (Separation Theorem) *Let $X \subseteq \mathbb{R}^n$ be a nonempty, closed and convex set and $z \in X^c$. Then there exist some $a \in \mathbb{R}^n \setminus \{0\}$ and $b \in \mathbb{R}$, such that for all $x \in X$ the inequalities*

$$\langle a, x \rangle \leq b < \langle a, z \rangle$$

are fulfilled.

Fig. 3.7 Alternatives in
Gordan's lemma

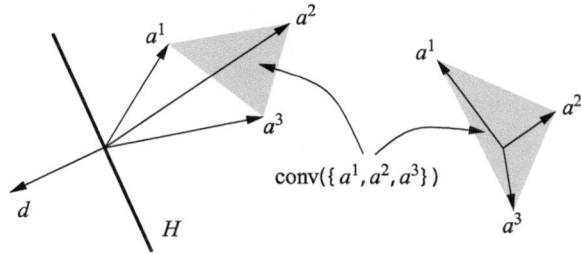

conv($\{a^1, a^2, a^3\}$)

For $a \in \mathbb{R}^n \setminus \{0\}$ and $b \in \mathbb{R}$ the set $H = \{x \in \mathbb{R}^n | \langle a, x \rangle = b\}$ is a *hyperplane*, which divides the space $\mathbb{R}^n$ into the two *half-spaces*

$$H_\leq := \{x \in \mathbb{R}^n | \langle a, x \rangle \leq b\} \quad \text{and} \quad H_> := \{x \in \mathbb{R}^n | \langle a, x \rangle > b\}.$$

Theorem 3.2.14 thus guarantees the existence of a hyperplane H with $X \subseteq H_\leq$ and $z \in H_>$, i.e., H *separates* the point z from the set X. It is easy to construct examples of a point and a *non*convex set that cannot be separated by a hyperplane.

We are now able to provide the second part of the proof for Gordan's lemma.

Proof of Theorem 3.2.13 (Part 2) We show that the negation of statement b implies statement a. To this end, let $0 \notin \mathrm{conv}(\{a^1, \ldots, a^r\})$. With the definitions $X := \mathrm{conv}(\{a^1, \ldots, a^r\})$ and $z := 0$ the assumptions of Theorem 3.2.14 are fulfilled (the proof that this set X is nonempty, closed and convex is left as an exercise for the reader), so there exist some $d \in \mathbb{R}^n$ and $b \in \mathbb{R}$ with

$$\forall x \in X : \quad \langle d, x \rangle \leq b < \langle d, 0 \rangle = 0.$$

Since the vectors $a^1, \ldots, a^r$ are particular elements of X, it follows $\langle d, a^k \rangle < 0$, $1 \leq k \leq r$, thus statement a.   □

Therefore, in simple terms, Gordan's lemma also states that the vectors $a^1, \ldots, a^r$ either strictly lie on one side of a hyperplane $H = \{x \in \mathbb{R}^n | \langle d, x \rangle = 0\}$ through the origin, or otherwise the origin is contained in the convex hull of the vectors $a^1, \ldots, a^r$ (Fig. 3.7).

We will use Gordan's lemma to characterize the condition from Corollary 3.2.4a. In order to also handle the condition in Corollary 3.2.4b, in the following result only one of the inequalities in statement a is strict, while all the others are nonstrict. The *conical hull* cone(A) of a set A, which is required to formulate the result, differs from the convex hull in that the weights λ_i do not need to sum up to one:

$$\mathrm{cone}(A) = \left\{ \sum_{i=1}^s \lambda_i a^i \,\middle|\, a^i \in A, \ \lambda_i \geq 0, \ 1 \leq i \leq s, \ s \in \mathbb{N} \right\}.$$

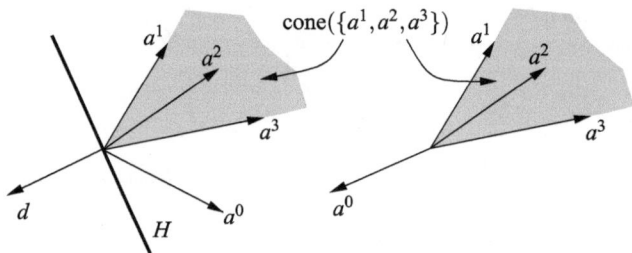

Fig. 3.8 Alternatives in Farkas' lemma

Theorem 3.2.15 (Farkas' Lemma) *For vectors $a^k \in \mathbb{R}^n$, $0 \leq k \leq r$, with $r \in \mathbb{N}$, exactly one of the following alternatives applies.*

(a) *The system $\langle a^0, d \rangle < 0$, $\langle a^k, d \rangle \leq 0$, $1 \leq k \leq r$, possesses a solution $d \in \mathbb{R}^n$.*
(b) *It holds $-a^0 \in \text{cone}(\{a^1, \dots, a^r\})$.*

Since the proof of Farkas' lemma not only uses the separation theorem (Theorem 3.2.14), but also a preliminary consideration from its derivation (Remark 3.2.34), we also postpone it to Sect. 3.2.5. Figure 3.8 illustrates the alternatives in Farkas' lemma.

The next result allows, among other things, to specify upper bounds for the number of vectors a^k (or, rather, of the positive weights λ_k) needed in the representations of the origin and of the vector $-a^0$ in Theorem 3.2.13b and Theorem 3.2.15b, respectively.

Theorem 3.2.16 (Carathéodory's Theorem) *For every set $A \subseteq \mathbb{R}^n$ the following statements are true.*

(a) *For each $\bar{x} \in \text{cone}(A) \setminus \{0\}$ there exist some $r \leq n$ and linearly independent vectors $x^k \in A$ as well as $\lambda_k > 0$, $1 \leq k \leq r$, with $\bar{x} = \sum_{k=1}^{r} \lambda_k x^k$.*
(b) *For each $\bar{x} \in \text{conv}(A)$ there exist some $r \leq n+1$ and $x^1, \dots, x^r \in A$, such that the vectors $x^2 - x^1, \dots, x^r - x^1$ are linearly independent and that $\bar{x} \in \text{conv}(\{x^1, \dots, x^r\})$ holds.*

Proof For the proof of statement a, let $\bar{x} = \sum_{k=1}^{r} \lambda_k x^k$ with $\lambda_k \geq 0$, $1 \leq k \leq r$, and linearly *dependent* $x^k \in A$, $1 \leq k \leq r$. The latter implies the existence of some $\mu_1, \ldots, \mu_r \in \mathbb{R}$, not all zero, with $0 = \sum_{k=1}^{r} \mu_k x^k$. For every $t \in \mathbb{R}$ we obtain

$$\bar{x} = \sum_{k=1}^{r} (\lambda_k + t\mu_k) x^k.$$

Without loss of generality, let $\mu_1, \ldots, \mu_s \neq 0$, $\mu_{s+1}, \ldots, \mu_r = 0$. We choose some $k_0 \in \{1, \ldots, s\}$ with

$$\forall 1 \leq k \leq s : \quad \left| \frac{\lambda_k}{\mu_k} \right| \geq \left| \frac{\lambda_{k_0}}{\mu_{k_0}} \right|$$

and set $t := -\lambda_{k_0}/\mu_{k_0}$.

Again without loss of generality, let $k_0 = s$. Then for all $1 \leq k \leq s - 1$

$$\lambda_k + t\mu_k = \lambda_k - \frac{\lambda_s}{\mu_s} \mu_k \geq \lambda_k - \left| \frac{\lambda_s}{\mu_s} \mu_k \right| = \lambda_k - \left| \frac{\lambda_s}{\mu_s} \right| \cdot |\mu_k| \geq \lambda_k - \left| \frac{\lambda_k}{\mu_k} \right| \cdot |\mu_k| = 0$$

holds, where the last equality follows from $\lambda_k \geq 0$. Overall, we obtain

$$\forall 1 \leq k \leq r : \quad \lambda_k + t\mu_k \geq 0 \quad \text{and} \quad \lambda_s + t\mu_s = 0,$$

so that the vector x^s is superfluous for the representation of $\bar{x}$.

Therefore, if the representation $\bar{x} = \sum_{k=1}^{r} \lambda_k x^k$ uses linearly dependent vectors x^k, some of them can be eliminated until only linearly *in*dependent vectors are left in the representation. In particular, then $r \leq n$ must hold.

To prove statement b, we note that $\bar{x} \in \text{conv}(A)$ holds if and only if $s \in \mathbb{N}$, $\lambda_k \geq 0$, $1 \leq k \leq s$, and

$$\begin{pmatrix} x^k \\ 1 \end{pmatrix} \in A \times \{1\} \subseteq \mathbb{R}^{n+1} \quad \text{with} \quad \begin{pmatrix} \bar{x} \\ 1 \end{pmatrix} = \sum_{k=1}^{s} \lambda^k \begin{pmatrix} x^k \\ 1 \end{pmatrix}$$

exist. This is equivalent to

$$\begin{pmatrix} \bar{x} \\ 1 \end{pmatrix} \in \text{cone}(A \times \{1\}) \setminus \{0\}.$$

According to statement a, there exist linearly independent vectors

$$\begin{pmatrix} x^k \\ 1 \end{pmatrix}, \ 1 \leq k \leq r,$$

and $\lambda_k > 0$ with

$$\begin{pmatrix} \bar{x} \\ 1 \end{pmatrix} = \sum_{k=1}^{r} \lambda^k \begin{pmatrix} x^k \\ 1 \end{pmatrix}.$$

It follows $\bar{x} \in \text{conv}(\{x^1, \dots, x^r\})$, and that the vectors $x^2 - x^1, \dots, x^r - x^1$ are linearly independent. The latter can be seen, for example, by elementary column transformation for the matrix

$$\begin{pmatrix} x^1 & \dots & x^r \\ 1 & \dots & 1 \end{pmatrix}.$$

In particular, this implies $r \leq n + 1$.   □

The convex hull $\text{conv}(\{x^1, \dots, x^r\})$ of r points with linearly independent vectors $x^2 - x^1, \dots, x^r - x^1$, as it appears in Theorem 3.2.16b, is called $(r - 1)$-simplex, and the vectors $x^1, \dots, x^r$ are then called *affinely independent*.

Carathéodory's theorem allows the following improvements of the statements in Gordan's lemma and Farkas' lemma, where $|A|$ denotes the number of elements of a set A.

Corollary 3.2.17

(a) *In Theorem 3.2.13b, weights λ_k with $|\{1 \leq k \leq r \mid \lambda_k > 0\}| \leq n + 1$ can be chosen.*
(b) *In Theorem 3.2.15b, weights λ_k with $|\{1 \leq k \leq r \mid \lambda_k > 0\}| \leq n$ can be chosen.*

3.2.4 First-Order Optimality Conditions Without Equality Constraints

We are now able to transform the first-order necessary optimality conditions from Corollary 3.2.4 into a form that can be handled algorithmically.

Theorem 3.2.18 (Fritz John's Theorem for $J = \emptyset$) *Let $\bar{x}$ be a local minimal point of P, at which the functions f and g_i, $i \in I_0(\bar{x})$, are differentiable. Then there exist multipliers $\kappa \geq 0$, $\lambda_i \geq 0$, $i \in I_0(\bar{x})$, not all zero, with*

$$\kappa \nabla f(\bar{x}) + \sum_{i \in I_0(\bar{x})} \lambda_i \nabla g_i(\bar{x}) = 0. \tag{3.3}$$

The multipliers κ and λ_i, $i \in I_0(\bar{x})$, can be chosen such that either $\kappa > 0$ and $|\{i \in I_0(\bar{x}) \mid \lambda_i > 0\}| \leq n$ or $\kappa = 0$ and $|\{i \in I_0(\bar{x}) \mid \lambda_i > 0\}| \leq n + 1$ holds.

Proof According to Corollary 3.2.4a, the system

$$\langle \nabla f(\bar{x}), d \rangle < 0, \quad \langle \nabla g_i(\bar{x}), d \rangle < 0, \ i \in I_0(\bar{x}),$$

is unsolvable, and according to Gordan's lemma (Theorem 3.2.13) this is equivalent to

$$0 \in \mathrm{conv}(\{ \nabla f(\bar{x}), \nabla g_i(\bar{x}), \ i \in I_0(\bar{x}) \}),$$

i.e., there exist $\kappa \geq 0$, $\lambda_i \geq 0$, $i \in I_0(\bar{x})$, with $\kappa + \sum_{i \in I_0(\bar{x})} \lambda_i = 1$ and

$$0 = \kappa \nabla f(\bar{x}) + \sum_{i \in I_0(\bar{x})} \lambda_i \nabla g_i(\bar{x}).$$

This is equivalent to the first assertion, because the nonnegative $\kappa, \lambda_i, \ i \in I_0(\bar{x})$, cannot simultaneously vanish if they sum up to one. On the other hand, the sum of every nonnegative and not simultaneously vanishing $\kappa, \ \lambda_i, \ i \in I_0(\bar{x})$, can be normalized to one by dividing each multiplier by their total sum.

The second assertion follows from Corollary 3.2.17. □

The limitation of the number of positive multipliers in the second assertion of Theorem 3.2.18 implies that the a priori unknown index set $I_0(\bar{x})$ is not an arbitrary subset of $I = \{1, \ldots, p\}$, which would lead to a case distinction in 2^p cases for the choice of $I_0(\bar{x})$, but that only sets $I_0(\bar{x}) \subseteq I$ with $|I_0(\bar{x})| \leq n + 1$ need to be considered. For $p > n + 1$ this can lead to a significant reduction in computational effort.

Example 3.2.19 For the parametric optimization problem $P(t)$ with objective function $f(x) = x$ and the single inequality constraint $g_1(t, x) = x^2 - t^2 \leq 0$, the feasible set is $M(t) = [-|t|, |t|]$, and $t \in \mathbb{R}$ denotes an exogenous parameter. We try to find the minimal point, which is geometrically obvious for every $t \in \mathbb{R}$ in this simple case, through formal considerations that can be transferred to geometrically more complex cases.

First, the Weierstrass theorem guarantees the existence of a global minimal point $\bar{x}(t)$ for every $t \in \mathbb{R}$. To find such a point, we look for conditions that $\bar{x}(t)$ must necessarily fulfill. First, a global minimal point is necessarily also a local minimal point. Therefore, it necessarily fulfills the condition from Theorem 3.2.18. Here, for the different possibilities of active index sets $I_0(\bar{x}(t))$, two cases need to be distinguished.

Case 1: $I_0(\bar{x}(t)) = \emptyset$. In this case, the contradiction $0 = f'(\bar{x}(t)) = 1$ would have to hold, so $\bar{x}(t)$ cannot fulfill $I_0(\bar{x}(t)) = \emptyset$.

Case 2: $I_0(\bar{x}(t)) = \{1\}$. The activity of the inequality means $0 = g_1(t, \bar{x}(t)) = \bar{x}^2 - t^2$, thus we obtain the two candidates $\bar{x}^1(t) = -|t|$ and $\bar{x}^2(t) = |t|$. The equation

$$0 = \kappa f'(\bar{x}^1(t)) + \lambda_1 g_1'(t, \bar{x}^1(t)) = \kappa - 2\lambda_1 |t|$$

is solvable with $(\kappa(t), \lambda_1(t)) = (2|t|, 1) \geq 0$, while

$$0 = \kappa f'(\bar{x}^2(t)) + \lambda_1 g_1'(t, \bar{x}^2(t)) = \kappa + 2\lambda_1 |t|$$

is solvable with nonvanishing $\kappa, \lambda_1 \geq 0$ only for $t = 0$ (g_1' here denotes the derivative of g_1 with respect to x). The only candidate for a local minimal point of $P(t)$ remains $\bar{x}^1(t) = -|t|$, so that $\bar{x}(t) = -|t|$ must be the unique global minimal point of $P(t)$ for every $t \in \mathbb{R}$.

Example 3.2.20 To motivate the MFCQ in Definition 3.2.3 we had pointed out that the unsolvability of the system used in the proof of Theorem 3.2.18,

$$\langle \nabla f(\bar{x}), d \rangle < 0, \quad \langle \nabla g_i(\bar{x}), d \rangle < 0, \ i \in I_0(\bar{x}),$$

can occur trivially, namely when $L_<(\bar{x}, M) = \emptyset$ holds. Indeed, in the case $t = 0$ in Example 3.2.19 we have $g_1(0, x) = x^2$, $M(0) = \{0\}$, and $\bar{x}(0) = 0$ is a global minimal point with $L_<(\bar{x}(0), M) = \emptyset$. From $g_1(0, \bar{x}(0)) = 0$ follows $I_0(\bar{x}) = \{1\}$, and Theorem 3.2.18 provides as a necessary optimality condition the existence of $\kappa, \lambda_1 \geq 0$, not both zero, with

$$0 = \kappa f'(\bar{x}(0)) + \lambda_1 g_1'(0, \bar{x}(0)) = \kappa \cdot 1 + \lambda_1 \cdot 0.$$

This enforces $\kappa(0) = 0$, while λ_1 can be any positive number, for example $\lambda_1(0) = 1$.

In situations where, as in Example 3.2.20, the multiplier κ vanishes, it is inconvenient that the gradient of the objective function f does not enter into the optimality condition (3.3), but only the constraints do. Fortunately, these situations can be characterized in a simple way.

Lemma 3.2.21 *Let $\bar{x}$ be a local minimal point of P, at which the functions f and g_i, $i \in I_0(\bar{x})$, are differentiable. Then (3.3) is solvable with $\kappa = 0$ and $\lambda_i \geq 0$, $i \in I_0(\bar{x})$, not all zero, if and only if the MFCQ is violated at $\bar{x}$.*

Proof The solvability of (3.3) with $\kappa = 0$ and $\lambda_i \geq 0$, $i \in I_0(\bar{x})$, not all zero, means

$$\sum_{i \in I_0(\bar{x})} \lambda_i \nabla g_i(\bar{x}) = 0$$

and is, hence, equivalent to $0 \in \text{conv}(\{\nabla g_i(\bar{x}), i \in I_0(\bar{x})\})$. Gordan's lemma provides the equivalence of the latter to the unsolvability of

$$\langle \nabla g_i(\bar{x}), d \rangle < 0, \ i \in I_0(\bar{x}),$$

which is the violation of the MFCQ at $\bar{x}$. □

We conclude the Karush-Kuhn-Tucker theorem, which is central for algorithms in constrained nonlinear optimization.

> **Theorem 3.2.22 (Karush-Kuhn-Tucker Theorem for $J = \emptyset$ Under MFCQ)** *Let $\bar{x}$ be a local minimal point of P, at which the functions f and g_i, $i \in I_0(\bar{x})$, are differentiable and at which the MFCQ holds. Then there exist multipliers $\lambda_i \geq 0$, $i \in I_0(\bar{x})$, with*
>
> $$\nabla f(\bar{x}) + \sum_{i \in I_0(\bar{x})} \lambda_i \nabla g_i(\bar{x}) = 0. \tag{3.4}$$
>
> *The λ_i can be chosen so that $|\{i \in I_0(\bar{x})| \lambda_i > 0\}| \leq n$ holds.*

Proof According to Theorem 3.2.18 and Lemma 3.2.21, (3.3) possesses a solution with $\kappa > 0$. After dividing (3.3) by κ, the first assertion follows with the new multipliers $\widetilde{\lambda}_i := \lambda_i/\kappa$. The second assertion follows from the second assertion in Theorem 3.2.18. □

Analogous to the remark after Theorem 3.2.18, the limitation of the number of positive multipliers in the second assertion of Theorem 3.2.22 implies that for the choice of $I_0(\bar{x})$ not 2^p cases need to be examined, but only sets $I_0(\bar{x}) \subseteq I$ with $|I_0(\bar{x})| \leq n$. In contrast to the Fritz-John condition from Theorem 3.2.18, this can already lead to a significant reduction in computational effort for $p > n$.

Although Theorem 3.2.22 is often used with the MFCQ or the even stronger LICQ (Definition 3.2.43), it is sometimes also useful to be able to weaken the assumption of the MFCQ to the ACQ.

> **Theorem 3.2.23 (Karush-Kuhn-Tucker Theorem for $J = \emptyset$ Under ACQ)** *The statement of Theorem 3.2.22 remains correct if there 'MFCQ' is replaced by 'ACQ'.*

Proof According to Corollary 3.2.4b, the system

$$\langle \nabla f(\bar{x}), d \rangle \; < \; 0, \quad \langle \nabla g_i(\bar{x}), d \rangle \; \leq \; 0, \; i \in I_0(\bar{x}),$$

is unsolvable, and according to *Farkas' lemma* (Theorem 3.2.15) this is equivalent to

$$- \nabla f(\bar{x}) \; \in \; \mathrm{cone}(\{ \nabla g_i(\bar{x}), \; i \in I_0(\bar{x}) \}),$$

i.e., there exist $\lambda_i \geq 0, i \in I_0(\bar{x})$, with (3.4). The second assertion in Theorem 3.2.22 follows from Corollary 3.2.17b. $\square$

The following result provides an important application of Theorem 3.2.23.

Corollary 3.2.24 *Let $g_i(x) = a_i^\mathsf{T} x + b_i$, $1 \leq i \leq p$, and let $\bar{x}$ be a local minimal point of P, at which f is differentiable. Then there exist multipliers $\lambda_i \geq 0$, $i \in I_0(\bar{x})$, with*

$$\nabla f(\bar{x}) + \sum_{i \in I_0(\bar{x})} \lambda_i \, a_i \; = \; 0.$$

The λ_i can be chosen so that $|\{ i \in I_0(\bar{x}) | \lambda_i > 0 \}| \leq n$ holds.

Proof Example 3.2.10 and Theorem 3.2.23. $\square$

We point out that we have derived the Karush-Kuhn-Tucker theorem for $J = \emptyset$ under the MFCQ in two independent ways. Both are based on Theorem 3.2.2. The first derivation showed the Fritz John theorem (Theorem 3.2.18) via Theorem 3.1.24 and Gordan's lemma and then derived the Karush-Kuhn-Tucker theorem (Theorem 3.2.22) under the MFCQ by a *second* application of Gordan's lemma. The alternative derivation first shows the Karush-Kuhn-Tucker theorem under the ACQ (Theorem 3.2.23) with the help of Farkas' lemma. Since the MFCQ is stronger than the ACQ (Theorem 3.2.8), this implies the Karush-Kuhn-Tucker theorem under the MFCQ.

In the literature, one almost exclusively finds this second derivation, justified in the historical development of nonlinear from linear optimization, in which the ACQ is automatically fulfilled (Example 3.2.10). However, the first derivation offers the advantage of also being transferable to cases of nonsmooth problems in which the MFCQ is not necessarily stronger than the ACQ. This effect occurs, for example, in generalized semi-infinite optimization (for details see [31, 32]).

Example 3.2.25 In the trust-region subproblem

$$TR^k : \qquad \min_{d \in \mathbb{R}^n} \; f(x^k) + \langle \nabla f(x^k), d \rangle + \tfrac{1}{2} d^\mathsf{T} A^k d \quad \text{s.t.} \quad \|d\|_2 \leq t^k$$

from Sect. 2.2.10, d denotes the decision variable, x^k the current iterate, A^k a symmetric, but not necessarily positive definite approximation of the Hessian matrix $D^2 f(x^k)$, and $t^k > 0$ the current search radius. Using the necessary optimality conditions, we obtain additional information about the search direction d^k determined as the minimal point of $T R^k$.

The feasible set of $T R^k$ is nonempty and compact as a ball, and the objective function $m^k(d) = f(x^k) + \langle \nabla f(x^k), d \rangle + \frac{1}{2} d^\mathsf{T} A^k d$ is continuous as a quadratic function. Therefore, the Weierstrass theorem guarantees the existence of a global minimal point d^k. Since $T R^k$ has no equality constraints, the necessary optimality conditions of this section seem to be applicable. However, to avoid a discussion of the differentiability of the inequality constraint at d^k, we equivalently reformulate it to $\|d\|_2^2 \le (t^k)^2$ and define the everywhere continuously differentiable inequality constraint function $g_1(d) := \|d\|_2^2 - (t^k)^2$. The index set of the inequality constraints is $I = \{1\}$.

In the next step, we note that at the minimal point d^k the MFCQ certainly holds, because in the case $I_0(d^k) = \emptyset$ there is nothing to show, and in the case $I_0(d^k) = \{1\}$, $\|d^k\|_2^2 = (t^k)^2 > 0$ and therefore

$$\langle \nabla g_1(d^k), -d^k \rangle = \langle 2d^k, -d^k \rangle = -2\|d^k\|_2^2 < 0$$

are true. According to Theorem 3.2.22, d^k thus fulfills one of the following two conditions, depending on the activity of the inequality constraint.

Case 1: $I_0(d^k) = \emptyset$. In this case, $\|d^k\|_2 < t^k$ and $0 = \nabla m^k(d^k) = \nabla f(x^k) + A^k d^k$ hold.

For $A^k \succ 0$ this results in $d^k = -\nabla_{A^k} f(x^k)$.

Case 2: $I_0(d^k) = \{1\}$. We obtain $\|d^k\|_2 = t^k$ and with some $\lambda_1 \ge 0$

$$0 = \nabla m^k(d^k) + \lambda_1 \nabla g_1(d^k) = \nabla f(x^k) + A^k d^k + 2\lambda_1 d^k. \tag{3.5}$$

Multiplying this inequality by d^k yields

$$0 = \langle \nabla f(x^k), d^k \rangle + (d^k)^\mathsf{T} A^k d^k + 2\lambda_1 (t^k)^2,$$

and solving for λ_1 gives

$$\lambda_1 = -\frac{\langle \nabla f(x^k), d^k \rangle + (d^k)^\mathsf{T} A^k d^k}{2(t^k)^2}.$$

Because of $\lambda_1 \ge 0$, d^k must necessarily satisfy the inequality $\langle \nabla f(x^k), d^k \rangle + (d^k)^\mathsf{T} A^k d^k \le 0$.

If the expression $2\lambda_1 = -(\langle \nabla f(x^k), d^k \rangle + (d^k)^\mathsf{T} A^k d^k)/(t^k)^2$ is so large that the matrix $A^k + 2\lambda_1 I$ is positive definite, then, according to (3.5), for the choice $\sigma^k =$

$2\lambda_1$ it matches with the search direction $d^k = -(A^k + \sigma^k I)^{-1}\nabla f(x^k)$ from the Levenberg-Marquardt approach (Exercise 2.2.49). For this reason, the Levenberg-Marquardt method is considered a precursor to trust-region methods.

3.2.5 Separation Theorem

The proof of the separation theorem (Theorem 3.2.14), which so far is missing for the verification of the results in Sects. 3.2.3 and 3.2.4, is based on the concept of orthogonal projection. For a set $X \subseteq \mathbb{R}^n$ and a point $z \in \mathbb{R}^n$, we consider the *projection problem*

$$Pr(z, X): \qquad \min \|x - z\|_2 \quad \text{s.t.} \quad x \in X,$$

whose global minimal points are those points from the set X that possess minimal Euclidean distance from z. The solvability of $Pr(z, X)$ is already guaranteed under mild assumptions.

Lemma 3.2.26 *Let $X \subseteq \mathbb{R}^n$ be a nonempty closed set and $z \in \mathbb{R}^n$. Then the projection problem $Pr(z, X)$ is solvable.*

Proof We choose a point $\bar{x} \in X$ and set $\alpha := \|\bar{x} - z\|_2$. For the objective function $f(x) := \|x - z\|_2$ of $Pr(z, X)$, the lower level set

$$f_{\leq}^{\alpha} = \{x \in \mathbb{R}^n \mid \|x - z\|_2 \leq \alpha\}$$

is a ball with center z and radius α, thus compact. Since the point $\bar{x}$ lies in both $f_{\leq}^{\alpha}$ and X, the set $\text{lev}_{\leq}^{\alpha}(f, X) = f_{\leq}^{\alpha} \cap X$ is nonempty. As the intersection of a compact and a closed set, it is also compact. Therefore, the strengthened Weierstrass theorem (Theorem 1.2.13) guarantees the solvability of $Pr(z, X)$. $\square$

According to Exercise 1.3.4, $Pr(z, X)$ has the same minimal points as

$$Pr^2(z, X): \qquad \min \|x - z\|_2^2 \quad \text{s.t.} \quad x \in X,$$

where the quadratic objective function of $Pr^2(z, X)$ has a positive definite Hessian matrix. According to [36], it is therefore strictly convex and can have at most one global minimal point on convex sets. For our application to the proof of the separation theorem, we will fortunately assume the set X in the projection problem $Pr(z, X)$ to be convex anyway, so that the following strengthening of Lemma 3.2.26 is shown for additionally convex sets $X \subseteq \mathbb{R}^n$.

Theorem 3.2.27 *Let $X \subseteq \mathbb{R}^n$ be a nonempty, closed and convex set as well as $z \in \mathbb{R}^n$. Then the problem $Pr(z, X)$ has a unique global minimal point.*

The unique global minimal point of $Pr(z, X)$ is called *orthogonal projection* of z onto X, shortly $pr(z, X)$, and the associated optimal *value* of $Pr(z, X)$ is the *distance* from z to X, shortly $dist(z, X)$ (for the connection to the parallel projection from Definition 1.3.6 see [36]). Note that, while Lemma 3.2.26 even applies to every norm, Theorem 3.2.27 does not (e.g., not for $\|\cdot\|_1$ or $\|\cdot\|_\infty$).

To the problem $Pr^2(z, X)$ we will apply the following result, which we can prove for every optimization problem with continuously differentiable convex objective function and convex feasible set. Its statement b *characterizes* global minimality by a stationarity condition, in which remarkably only the objective function, but *not* the feasible set are linearized.

Theorem 3.2.28 (Variational Formulation of Convex Problems) *Let the set $M \subseteq \mathbb{R}^n$ and the function $f \in C^1(M, \mathbb{R})$ be convex. Then the following statements hold.*

(a) *The point $\bar{x} \in \mathbb{R}^n$ a global minimal point of*

$$P: \quad \min f(x) \quad \text{s.t.} \quad x \in M,$$

if and only if $\bar{x}$ is a global minimal point of

$$\widetilde{P}_{\text{lin}}(\bar{x}): \quad \min_x \langle \nabla f(\bar{x}), x - \bar{x} \rangle \quad \text{s.t.} \quad x \in M.$$

(b) *The set of global minimal points of P coincides with the set*

$$\{\bar{x} \in M | \langle \nabla f(\bar{x}), x - \bar{x} \rangle \geq 0 \text{ for all } x \in M\}.$$

Proof For the proof of statement a, let $\bar{x}$ be a global minimal point of $\widetilde{P}_{\text{lin}}(\bar{x})$. Then in particular $\bar{x} \in M$ holds, so that $\bar{x}$ is also feasible for P. Furthermore, according to Theorem 2.1.40 for all $x \in M$

$$f(x) \geq f(\bar{x}) + \langle \nabla f(\bar{x}), x - \bar{x} \rangle \geq f(\bar{x}) + \langle \nabla f(\bar{x}), \bar{x} - \bar{x} \rangle = f(\bar{x})$$

is true, so $\bar{x}$ is a global minimal point of P.

On the other hand, let $\bar{x}$ be a global minimal point of P. In particular, $\bar{x}$ is then feasible for $\widetilde{P}_{\text{lin}}(\bar{x})$. Let us assume that $\bar{x}$ is *not* a global minimal point of $\widetilde{P}_{\text{lin}}(\bar{x})$. Then there exists some $x \in M$ with

$$\langle \nabla f(\bar{x}), x - \bar{x} \rangle \; < \; \langle \nabla f(\bar{x}), \bar{x} - \bar{x} \rangle \; = \; 0.$$

The direction $d := x - \bar{x}$ is therefore a first-order descent direction for f in $\bar{x}$ in the sense of Definition 2.1.7. Moreover, points on the half-line from $\bar{x}$ along d do not leave M for sufficiently small step sizes $t > 0$, because due to $\bar{x}, x \in M$ and the convexity of M it follows for all step sizes $t \in [0, 1]$

$$\bar{x} + td \; = \; (1 - t)\bar{x} + tx \; \in \; M.$$

Thus, d is a feasible descent direction for P in $\bar{x}$ in the sense of Definition 3.1.4, which contradicts the minimality of $\bar{x}$ for P according to Exercise 3.1.5.

To see statement b, observe that a point $\bar{x} \in \mathbb{R}^n$ is a global minimal point of $\widetilde{P}_{\text{lin}}(\bar{x})$ if and only if the conditions $\bar{x} \in M$ and

$$\langle \nabla f(\bar{x}), x - \bar{x} \rangle \; \geq \; \langle \nabla f(\bar{x}), \bar{x} - \bar{x} \rangle \quad \text{for all } x \in M$$

apply. From $\langle \nabla f(\bar{x}), \bar{x} - \bar{x} \rangle = 0$ and statement a, the assertion follows. □

Theorem 3.2.28b motivates the introduction of the following concepts.

Definition 3.2.29 (Polar Cone) For a set $A \subseteq \mathbb{R}^n$ the set

$$A^\circ \; = \; \{s \in \mathbb{R}^n \,|\, \langle s, d \rangle \leq 0 \;\text{ for all } d \in A\}$$

is called the *polar cone* of A.

Exercise 3.2.30 Show for each set $A \subseteq \mathbb{R}^n$ that its polar cone A° is a convex and closed cone with $0 \in A^\circ$.

Definition 3.2.31 (Normal Cone to Convex Sets) For a convex set $X \subseteq \mathbb{R}^n$ and $\bar{x} \in X$, the set

$$N(\bar{x}, X) \; := \; (X - \bar{x})^\circ \; = \; \{s \in \mathbb{R}^n \,|\, \langle s, x - \bar{x} \rangle \leq 0 \;\text{ for all } x \in X\}$$

is called the *normal cone* to X at $\bar{x}$. The elements s of the normal cone $N(\bar{x}, X)$ are called (outer) *normal directions* to X at $\bar{x}$.

Fig. 3.9 Normal cone to a convex set

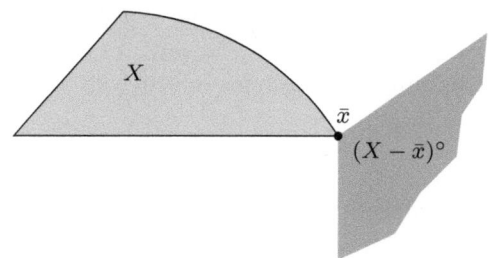

Geometrically interpreted, the normal cone to X at $\bar{x}$ contains exactly those directions $s \in \mathbb{R}^n$, with which no vector d forms an acute angle, for which $\bar{x} + d$ lies in X (which can be seen by the relation $x = \bar{x} + d$; Fig. 3.9). For a more detailed discussion of the normal cone, we refer to Sect. 3.2.6 and [33]. Here we rather continue the main line of arguments and reformulate the statement of Theorem 3.2.28b with the help of the normal cone.

Corollary 3.2.32 *Let the set $M \subseteq \mathbb{R}^n$ and the function $f \in C^1(M, \mathbb{R})$ be convex. Then $\bar{x} \in \mathbb{R}^n$ is a global minimal point of*

$$P : \quad \min f(x) \quad \text{s.t.} \quad x \in M,$$

if and only if the conditions $\bar{x} \in M$ and

$$- \nabla f(\bar{x}) \in N(\bar{x}, M)$$

are fulfilled.

In the next step, we apply Corollary 3.2.32 to the problem of orthogonal projection, as announced.

Theorem 3.2.33 (Projection Lemma) *Let $X \subseteq \mathbb{R}^n$ be a nonempty, closed and convex set and $z \in \mathbb{R}^n$. Then the unique minimal point $\mathrm{pr}(z, X)$ of the projection problem*

$$Pr(z, X) : \quad \min \|x - z\|_2 \quad \text{s.t.} \quad x \in X$$

is simultaneously the unique solution of the conditions

$$x \in X \quad \text{and} \quad z \in x + N(x, X).$$

Proof The unique solvability of $Pr(z, X)$ was already shown in Theorem 3.2.27. The unique minimal point $\bar{x} = \mathrm{pr}(z, X)$ of $Pr(z, X)$ is also the unique minimal point of the convex optimization problem with continuously differentiable objective function

$$Pr^2(z, X) : \qquad \min \|x - z\|_2^2 \quad \text{s.t.} \quad x \in X.$$

According to Corollary 3.2.32, this is exactly the case for $\bar{x} \in X$ and

$$-2(\bar{x} - z) \in N(\bar{x}, X).$$

Due to the cone property of $N(\bar{x}, X)$, the latter is equivalent to the assertion. $\square$

After these extensive preparations, we can prove the separation theorem.

Proof of Theorem 3.2.14 We consider the orthogonal projection $\bar{x} = \mathrm{pr}(z, X)$ of z onto X. According to Theorem 3.2.33, it fulfills $z \in \bar{x} + N(\bar{x}, X)$. Thus, for all $x \in X$ the inequality

$$\langle z - \bar{x}, x - \bar{x} \rangle \le 0$$

is true. With $a := z - \bar{x}$ and $b := \langle z - \bar{x}, \bar{x} \rangle$, it follows for all $x \in X$

$$\langle a, x \rangle - b = \langle z - \bar{x}, x \rangle - \langle z - \bar{x}, \bar{x} \rangle = \langle z - \bar{x}, x - \bar{x} \rangle \le 0 \qquad (3.6)$$

and

$$\langle a, z \rangle - b = \langle z - \bar{x}, z \rangle - \langle z - \bar{x}, \bar{x} \rangle = \|z - \bar{x}\|_2^2 > 0,$$

where both the strict positivity in the last inequality and the assertion $a \ne 0$ are guaranteed by $z \in X^c$ and $\bar{x} \in X$. $\square$

The evaluation of the inequality (3.6) at $\bar{x} \in X$ shows that the constructed hyperplane contains the point $\bar{x}$ and thus touches the set X. Indeed, we have not just constructed any separating hyperplane, but even a *supporting hyperplane* to the set X.

Remark 3.2.34 After conducting its proof using the concept of orthogonal projection, the assertion of the separation theorem (Theorem 3.2.14) can be extended to include the information that one may choose $a := z - \mathrm{pr}(z, X)$ and $b := \langle z - \mathrm{pr}(z, X), \mathrm{pr}(z, X) \rangle$.

With the help of this additional information, we now prove Farkas' lemma (Theorem 3.2.15).

Proof of Theorem 3.2.15 We show the equivalence of statement b and the negation of statement a. To this end, let $-a^0 \in \mathrm{cone}(\{a^1, \dots, a^r\})$, so there exist

$\lambda_1, \ldots, \lambda_r \geq 0$ with $-a^0 = \sum_{k=1}^r \lambda_k a^k$. Suppose statement a is true. Then there exists some $d \in \mathbb{R}^n$ with $\langle a^0, d \rangle < 0$, $\langle a^k, d \rangle \leq 0$, $1 \leq k \leq r$, and thus

$$0 < -\langle a^0, d \rangle = \sum_{k=1}^r \underbrace{\lambda_k}_{\geq 0} \underbrace{\langle a^k, d \rangle}_{\leq 0} \leq 0.$$

Since this is a contradiction, the negation of statement a holds.

To conclude statement a from the negation of statement b, let

$$-a^0 \notin \mathrm{cone}(\{a^1, \ldots, a^r\}) =: K.$$

Since K is nonempty, closed and convex (again left to the reader as an exercise), according to Theorem 3.2.14 there exist some $d \in \mathbb{R}^n \setminus \{0\}$ and $b \in \mathbb{R}$ with

$$\forall x \in K : \quad \langle d, x \rangle \leq b < \langle d, -a^0 \rangle, \tag{3.7}$$

where according to Remark 3.2.34 d can be chosen in the form $d = -a^0 - \mathrm{pr}(-a^0, K)$, and $b = \langle d, \mathrm{pr}(-a^0, K) \rangle$. We will show that this vector d solves the system of inequalities in statement a.

The crucial observation for this is that b must vanish. Indeed, from the cone property of K it follows that with $\mathrm{pr}(-a^0, K)$ also the vectors $2\mathrm{pr}(-a^0, K)$ and $(1/2)\mathrm{pr}(-a^0, K)$ are in K, so that the first inequality in (3.7) yields both

$$\langle d, 2\mathrm{pr}(-a^0, K) \rangle \leq \langle d, \mathrm{pr}(-a^0, K) \rangle$$

and

$$\left\langle d, \frac{1}{2}\mathrm{pr}(-a^0, K) \right\rangle \leq \langle d, \mathrm{pr}(-a^0, K) \rangle,$$

in total, $b = \langle d, \mathrm{pr}(-a^0, K) \rangle = 0$. Therefore, the first inequality in (3.7) implies

$$\forall x \in K : \quad \langle d, x \rangle \leq 0,$$

and in particular for the vectors $a^k \in K$ it follows $\langle d, a^k \rangle \leq 0$, $1 \leq k \leq r$. The strict inequality $\langle a^0, d \rangle < 0$ follows from $b = 0$ due to the second inequality in (3.7). This proves statement a. □

3.2.6 Normal Cones

The concept of the normal cone, introduced among other things for the proof of the separation theorem, also allows a new perspective on the stationarity condition

Fig. 3.10 Polar cone to a
nonconvex set

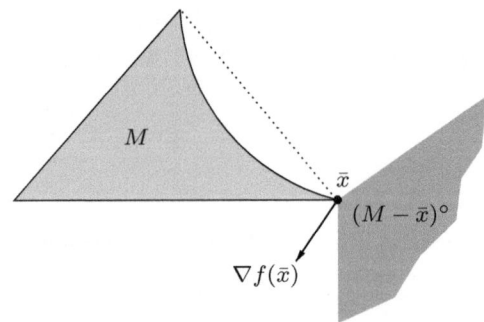

for constrained optimization problems. In this section we assume without further
mention that all involved functions are differentiable at the respective point $\bar{x}$.

According to Theorem 3.2.2, every local minimal point $\bar{x}$ of

$$P: \quad \min f(x) \quad \text{s.t.} \quad x \in M$$

is necessarily stationary in the sense of Definition 3.2.1, so it holds

$$\langle -\nabla f(\bar{x}), d \rangle \leq 0 \text{ for all } d \in C(\bar{x}, M). \tag{3.8}$$

For this result, neither the convexity of M nor of f on M is required. Motivated
by Corollary 3.2.32, our goal is to reformulate the stationarity condition in the
form $-\nabla f(\bar{x}) \in N(\bar{x}, M)$ with a suitably defined normal cone. Figure 3.10 shows
that our previous definition of the normal cone to *convex* sets (Definition 3.2.31) is
insufficient for optimization over nonconvex sets M, because although $\bar{x}$ is a local
minimal point of f over M, $-\nabla f(\bar{x})$ does not lie in the polar cone $(M - \bar{x})^\circ$.

The reason for this is that the definition of the normal cone to a convex set X may
be based on the *global* geometry of X, while for nonconvex sets X we need a *local*
construction. It is appropriate to model this local aspect through the outer tangent
cone to X at $\bar{x}$, and to use the polar cone of the latter as the normal cone.

Definition 3.2.35 (Normal Cone to Arbitrary Sets) For a set $X \subseteq \mathbb{R}^n$ and
$\bar{x} \in X$, we call

$$N(\bar{x}, X) = C^\circ(\bar{x}, X)$$

normal cone to X at $\bar{x}$. The elements s of the normal cone $N(\bar{x}, X)$ are again
called (outer) *normal directions* to X at $\bar{x}$.

For convex sets X, the two available definitions of normal cones fortunately coincide.

Theorem 3.2.36 *For every convex set $X \subseteq \mathbb{R}^n$ and $\bar{x} \in X$ it holds*

$$(X - \bar{x})^{\circ} = C^{\circ}(\bar{x}, X).$$

Proof To prove the inclusion $\subseteq$, let $s \in (X - \bar{x})^{\circ}$. Then $\langle s, x - \bar{x} \rangle \leq 0$ holds for all $x \in X$. For each $d \in C(\bar{x}, X)$ there also exist sequences $t^k \searrow 0$ and $d^k \to d$ with $\bar{x} + t^k d^k \in X, k \in \mathbb{N}$. From this it follows for all $k \in \mathbb{N}$

$$0 \geq \langle s, (\bar{x} + t^k d^k) - \bar{x} \rangle = t^k \langle s, d^k \rangle,$$

and because of $t^k > 0$ also $0 \geq \langle s, d^k \rangle$. In the limit $k \to \infty$ we thus obtain $0 \geq \langle s, d \rangle$, so that s lies in $C^{\circ}(\bar{x}, X)$.

To see the inclusion $\supseteq$, we choose some $s \in C^{\circ}(\bar{x}, X)$ and an arbitrary point $x \in X$. Then the direction $d := x - \bar{x}$ lies in $C(\bar{x}, X)$, because due to the convexity of X it holds $\bar{x} + td \in X$ for all $t \in [0, 1]$. In particular, we thus obtain with the choices $t^k := 1/k$ and $d^k := d$ the property $\bar{x} + t^k d^k \in X, k \in \mathbb{N}$. Due to $d \in C(\bar{x}, X)$ it holds by assumption

$$0 \geq \langle s, d \rangle = \langle s, x - \bar{x} \rangle,$$

thus $s \in (X - \bar{x})^{\circ}$. $\square$

Indeed, Theorem 3.2.2 states in this terminology that at a local minimal point $\bar{x}$ of f on M necessarily

$$- \nabla f(\bar{x}) \in N(\bar{x}, M)$$

is true, as this is a reformulation of the stationarity condition (3.8). The stationarity condition can thus also be interpreted geometrically as the vector $-\nabla f(\bar{x})$ (i.e., the steepest descent direction for f at $\bar{x}$) being an outer normal direction to M at $\bar{x}$ (Fig. 3.11).

Since the functional description of the set M plays no role for this result (but only its geometry), no constraint qualifications were required for its derivation. In the following, however, we will see how they help in deriving an *explicit representation* for the elements of the normal cone.

Exercise 3.2.37 Show for all $\bar{x} \in M$ the inclusion

$$L^{\circ}_{\leq}(\bar{x}, M) \subseteq N(\bar{x}, M).$$

Fig. 3.11 Normal cone to a
nonconvex set

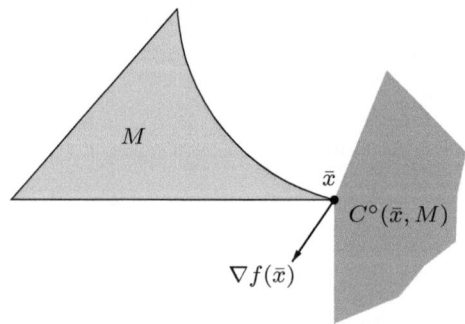

Exercise 3.2.38 Using Farkas' lemma (Theorem 3.2.15), show for all $\bar{x} \in M$ the
identity

$$L^\circ_\leq(\bar{x}, M) \;=\; \mathrm{cone}(\{\,\nabla g_i(\bar{x}),\; i \in I_0(\bar{x})\,\}).$$

Exercise 3.2.39 Let the ACQ hold at $\bar{x} \in M$. Show with the help of Exercises 3.2.37 and 3.2.38 the identity

$$N(\bar{x}, M) \;=\; \mathrm{cone}(\{\,\nabla g_i(\bar{x}),\; i \in I_0(\bar{x})\,\}).$$

For every local minimal point $\bar{x}$ of P, at which the ACQ is fulfilled, it thus holds

$$-\,\nabla f(\bar{x}) \;\in\; \mathrm{cone}(\{\,\nabla g_i(\bar{x}),\; i \in I_0(\bar{x})\,\}).$$

This is precisely the assertion of the Karush-Kuhn-Tucker theorem (Theorem 3.2.23). However, such an alternative derivation of the Karush-Kuhn-Tucker conditions does not cover the case where no constraint qualification is fulfilled at a local minimal point. How the Fritz-John condition results in this case, we have seen instead with the derivation from Sect. 3.2.4.

From the above derivation of the Karush-Kuhn-Tucker theorem, we can draw another important conclusion. Crucial for this is the identity from Exercise 3.2.39, which is equivalent to $C^\circ(\bar{x}, M) = L^\circ_\leq(\bar{x}, M)$. As seen there, the ACQ is *sufficient* for this identity, but there are also simple examples where it still holds while the ACQ is violated. Thus, we obtain an even weaker constraint qualification than the ACQ and say that the *Guignard constraint qualification (GCQ)* holds at $\bar{x}$ in M, if $C^\circ(\bar{x}, M) = L^\circ_\leq(\bar{x}, M)$ is fulfilled. Thus, in Theorem 3.2.22, the requirement of the MFCQ can be replaced not only by the ACQ, but even by the GCQ. This is actually the *weakest possible* constraint qualification, since in the case of its violation, there always exists some at $\bar{x}$ continuously differentiable objective function f that has a local minimal point there, while at the same time the Karush-Kuhn-Tucker conditions are violated [15].

The presented interpretation of the stationarity condition using the normal cone can, with some effort, be extended to several nonsmooth constrained optimization problems (see, for example, [28]).

3.2.7 First-Order Optimality Conditions with Equality Constraints

From now on, let the feasible set of P be described again by inequality *and* equality constraints, i.e., it holds

$$M = \{x \in \mathbb{R}^n | \, g_i(x) \leq 0, \; i \in I, \; h_j(x) = 0, \; j \in J\}.$$

To reduce this case to the already considered one without equality constraints, it may seem obvious to split each equality $h(x) = 0$ into two inequalities $g_1(x) := h(x) \leq 0$ and $g_2(x) := -h(x) \leq 0$. A significant disadvantage of this approach is that the new inequalities are active at each point $x \in M$ and have gradients $\nabla g_1(x) = \nabla h(x)$ and $\nabla g_2(x) = -\nabla h(x)$ oriented in opposite directions. Therefore, the MFCQ is violated everywhere in M, regardless of how regular the original description of M was. Such a transformation of an equality into two inequality constraints is only advisable for polyhedral sets M, because at least the validity of the ACQ is not affected there, even if the transformation destroys other regularity properties.

For nonlinear problems, one must therefore choose a different path to derive optimality conditions for equality-constrained problems. Various approaches exist for this [13, 16]. We choose the geometrically intuitive approach of solving the equations and only considering points x from their solution set. Since this is often not possible explicitly, one could rely instead on the implicit function theorem. Our following and equivalent approach consists of a nonlinear coordinate transformation, the treatment of which is based on the inverse function theorem.

Indeed, as illustrated in Fig. 3.12, by introducing a new coordinate system we will try to transform a set described by nonlinear equality constraints at least locally around one of its elements into a simpler set, preferably a linear space of the corresponding dimension. In the following, we will discuss in detail the construction of an appropriate nonlinear coordinate transformation.

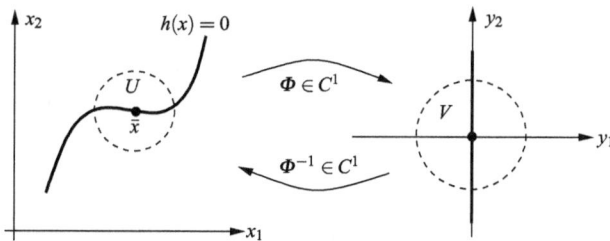

Fig. 3.12 Transformation to new coordinates

The transformation of the x-coordinates into y-coordinates is realized by a C^1-*diffeomorphism* Φ and setting $y = \Phi(x)$, where for $U, V \subseteq \mathbb{R}^n$ the function $\Phi : U \to V$ is called C^1-diffeomorphism, if and only if Φ is bijective and both Φ and the inverse function Φ^{-1} are continuously differentiable on their respective domain.

We consider a point $\bar{x} \in \mathbb{R}^n$ with $h(\bar{x}) = 0$ and, for simplicity, we assume the function $h : \mathbb{R}^n \to \mathbb{R}^q$ to be continuously differentiable on all of $\mathbb{R}^n$ (which could be restricted to continuous differentiability at $\bar{x}$ if necessary). The crucial assumption in the following will be *the linear independence of the gradients* $\nabla h_1(\bar{x}), \ldots, \nabla h_q(\bar{x})$. Because of $q < n$, then

$$\dim \ker Dh(\bar{x}) = \dim\{d \in \mathbb{R}^n \mid \langle \nabla h_j(\bar{x}), d \rangle = 0, \; j \in J\} = n - q > 0$$

holds, so that the kernel of the Jacobian matrix $Dh(\bar{x})$ is at least one-dimensional. We choose an orthonormal basis $\eta_{q+1}, \ldots, \eta_n \in \mathbb{R}^n$ of this kernel (i.e., the vectors $\eta_{q+1}, \ldots, \eta_n$ form a basis of the kernel, each have length one, and they are pairwise orthogonal) and define for the indices $q + 1 \leq j \leq n$ the auxiliary functions

$$h_j(x) = \langle \eta_j, x - \bar{x} \rangle.$$

These also fulfill $h_j(\bar{x}) = 0$, and they possess the gradients $\nabla h_j(\bar{x}) = \eta_j$, $q + 1 \leq j \leq n$. After this preliminary work, we define

$$\Phi(x) := \begin{pmatrix} h_1(x) \\ \vdots \\ h_q(x) \\ h_{q+1}(x) \\ \vdots \\ h_n(x) \end{pmatrix}.$$

This function maps from $\mathbb{R}^n$ to $\mathbb{R}^n$, is continuously differentiable, fulfills $\Phi(\bar{x}) = 0$, and its Jacobian

$$D\Phi(\bar{x}) := \begin{pmatrix} Dh_1(\bar{x}) \\ \vdots \\ Dh_q(\bar{x}) \\ \eta_{q+1}^{\mathsf{T}} \\ \vdots \\ \eta_n^{\mathsf{T}} \end{pmatrix}$$

is nonsingular by construction of the vectors η_j. The inverse functions theorem thus guarantees the existence of a neighborhood U of $\bar{x}$ and a neighborhood V of $0 \in \mathbb{R}^n$, such that Φ is a C^1-diffeomorphism between U and V.

Moreover, the inverse function theorem yields

$$D[\Phi^{-1}(0)] = (D\Phi(\bar{x}))^{-1}.$$

With this we obtain

$$x \in U \quad \Leftrightarrow \quad \exists\, y \in V : y = \Phi(x).$$

To understand the simplification of the zero set of h in the new coordinates indicated in Fig. 3.12, we consider the equivalences

$$x \in U, \, h(x) = 0 \Leftrightarrow \exists\, y \in V : y = \Phi(x) = \begin{pmatrix} h(x) \\ h_{q+1}(x) \\ \vdots \\ h_n(x) \end{pmatrix}, \, h(x) = 0$$

$$\Leftrightarrow \exists\, y \in V : y = \Phi(x), \, y_1 = \dots = y_q = 0$$

$$\Leftrightarrow \exists\, z \in W : x = \Phi^{-1}\begin{pmatrix} 0_q \\ z \end{pmatrix},$$

where

$$W = \left\{ z \in \mathbb{R}^{n-q} \,\middle|\, \begin{pmatrix} 0_q \\ z \end{pmatrix} \in V \right\}$$

is a neighborhood of $\bar{z} = 0$ in $\mathbb{R}^{n-q}$. Locally around $\bar{x}$, M can thus be described by the $n - q$ degrees of freedom of the vector z, using the C^1-diffeomorphism Φ. In this case, one says that M is a C^1-*manifold of dimension* $n - q$ locally around $\bar{x}$.

Next, we apply this transformation to a local minimal point $\bar{x}$ of P. By definition, $\bar{x}$ is a local minimal point of P if there exists another neighborhood U_2 of $\bar{x}$ such that all $x \in U_2 \cap M$ satisfy the inequality $f(x) \geq f(\bar{x})$. Without loss of generality, let $U \subseteq U_2$. Then in particular we have

$$\forall\, x \in U \text{ with } g(x) \leq 0, \, h(x) = 0 : \quad f(x) \geq f(\bar{x})$$

and thus

$$\forall\, z \in W \text{ with } \tilde{g}(z) \leq 0 : \quad \tilde{f}(z) \geq \tilde{f}(0), \tag{3.9}$$

where

$$\tilde{f}(z) := f\left(\Phi^{-1}\begin{pmatrix} 0_q \\ z \end{pmatrix}\right)$$

and

$$\widetilde{g}_i(z) := g_i\left(\Phi^{-1}\begin{pmatrix}0_q\\z\end{pmatrix}\right), \quad i \in I,$$

are the functions f and g_i 'in new coordinates'. As a composition of C^1-functions, these transformed functions are again continuously differentiable.

The statement in (3.9) means that $\bar{z} = 0$ is a local minimal point of the transformed problem

$$\widetilde{P}: \qquad \min_{z \in \mathbb{R}^{n-q}} \widetilde{f}(z) \quad \text{s.t.} \quad \widetilde{g}(z) \le 0.$$

The task of the coordinate transformation, to locally reduce the equality-constrained problem P to a problem without equality constraints, is thus fulfilled. We can therefore apply the results from Sect. 3.2.4 to the transformed problem $\widetilde{P}$, and then it remains to transform the results back to the original coordinates.

For example, according to Theorem 3.2.18, there exist multipliers $\kappa \ge 0$, $\lambda_i \ge 0$, $i \in I_0(\bar{z})$, not all zero, with

$$\kappa \nabla \widetilde{f}(\bar{z}) + \sum_{i \in I_0(\bar{z})} \lambda_i \nabla \widetilde{g}_i(\bar{z}) = 0, \tag{3.10}$$

where for $\kappa > 0$ the choice $|\{i \in I_0(\bar{z})|\ \lambda_i > 0\}| \le n - q$ is possible, and for $\kappa = 0$ the choice $|\{i \in I_0(\bar{z})|\ \lambda_i > 0\}| \le n - q + 1$.

We have

$$I_0(\bar{z}) = \left\{i \in I \,\middle|\, 0 = \widetilde{g}_i(\bar{z}) = g_i\left(\Phi^{-1}\begin{pmatrix}0_q\\\bar{z}\end{pmatrix}\right) = g_i(\Phi^{-1}(0_n)) = g_i(\bar{x})\right\}$$

$$= I_0(\bar{x}),$$

$$D\widetilde{f}(\bar{z}) = D\left[f\left(\Phi^{-1}\begin{pmatrix}0_q\\z\end{pmatrix}\right)\right]_{z=\bar{z}} = Df(\Phi^{-1}(0)) \cdot D(\Phi^{-1}(0)) \cdot \begin{pmatrix}0_{q\times(n-q)}\\I_{n-q}\end{pmatrix}$$

$$= Df(\bar{x}) \cdot (D\Phi(\bar{x}))^{-1} \cdot \begin{pmatrix}0\\I\end{pmatrix}$$

and analogously

$$D\widetilde{g}_i(\bar{z}) = Dg_i(\bar{x}) \cdot (D\Phi(\bar{x}))^{-1} \cdot \begin{pmatrix}0\\I\end{pmatrix}$$

for all $i \in I$.

The matrix

$$\Psi := (D\Phi(\bar{x}))^{-1} \cdot \begin{pmatrix} 0 \\ I \end{pmatrix}$$

appearing in these representations is the unique solution of the system

$$D\Phi(\bar{x}) \cdot \Psi = \begin{pmatrix} 0 \\ I \end{pmatrix}.$$

Since we have chosen the vectors $\eta_{q+1}, \ldots, \eta_n$ as an orthonormal basis, it is easy to verify that $\Psi = (\eta_{q+1}, \ldots, \eta_n)$ is the desired solution.

The Fritz-John condition (3.10) is therefore equivalent to

$$\left[\kappa\, Df(\bar{x}) + \sum_{i \in I_0(\bar{x})} \lambda_i\, Dg_i(\bar{x}) \right] \cdot \Psi = 0,$$

which is equivalent to

$$\kappa\, \nabla f(\bar{x}) + \sum_{i \in I_0(\bar{x})} \lambda_i\, \nabla g_i(\bar{x}) \in \ker \Psi^\mathsf{T} = \mathrm{img}(\nabla h_1(\bar{x}), \ldots, \nabla h_q(\bar{x}))$$

Therefore, there exist multipliers $\mu_1, \ldots, \mu_q \in \mathbb{R}$ with

$$\kappa\, \nabla f(\bar{x}) + \sum_{i \in I_0(\bar{x})} \lambda_i\, \nabla g_i(\bar{x}) + \sum_{j \in J} \mu_j\, \nabla h_j(\bar{x}) = 0.$$

This is the desired condition in original coordinates. We summarize.

Theorem 3.2.40 (Fritz John's Theorem) *Let $\bar{x}$ be a local minimal point of P, at which the functions f, g_i, $i \in I_0(\bar{x})$, and h_j, $j \in J$, are continuously differentiable. Then there exist multipliers $\kappa \geq 0$, $\lambda_i \geq 0$, $i \in I_0(\bar{x})$, $\mu_j \in \mathbb{R}$, $j \in J$, not all zero, with*

$$\kappa \nabla f(\bar{x}) + \sum_{i \in I_0(\bar{x})} \lambda_i\, \nabla g_i(\bar{x}) + \sum_{j \in J} \mu_j\, \nabla h_j(\bar{x}) = 0.$$

One can choose κ and the λ_i such that either $\kappa > 0$ and $|\{i \in I_0(\bar{x})|\, \lambda_i > 0\}| \leq n - q$ or $\kappa = 0$ and $|\{i \in I_0(\bar{x})|\, \lambda_i > 0\}| \leq n - q + 1$ holds.

Proof We have just proven the assertions in the case that the gradients $\nabla h_j(\bar{x}), j \in J$, are linearly independent. In this case, not all multipliers $\kappa \geq 0$, $\lambda_i \geq 0$, $i \in I_0(\bar{x})$, simultaneously vanish. If, on the other hand, the gradients $\nabla h_j(\bar{x}), j \in J$, are linearly dependent, then there exist $\mu_1, \ldots, \mu_q \in \mathbb{R}$, not all zero, with

$$\sum_{j \in J} \mu_j \nabla h_j(\bar{x}) = 0.$$

The claim then follows already by setting $\kappa = \lambda_i = 0$, $i \in I_0(\bar{x})$.   □

To formulate the Karush-Kuhn-Tucker theorem in the presence of equality constraints, we need a corresponding generalization of the MFCQ to equality-constrained problems.

Definition 3.2.41 (Mangasarian-Fromovitz Constraint Qualification)
The point $\bar{x} \in M$ satisfies the *Mangasarian-Fromovitz constraint qualification (MFCQ)*, if the following statements hold:

(a) The vectors $\nabla h_j(\bar{x}), j \in J$, are linearly independent.
(b) There exists some $d \in \mathbb{R}^n$ with

$$\langle \nabla g_i(\bar{x}), d \rangle < 0, \ i \in I_0(\bar{x}), \ \langle \nabla h_j(\bar{x}), d \rangle = 0, \ j \in J.$$

Theorem 3.2.42 (Karush-Kuhn-Tucker Theorem) *Let $\bar{x}$ be a local minimal point of P, at which the functions f, g_i, $i \in I_0(\bar{x})$, and h_j, $j \in J$, are continuously differentiable, and at which the MFCQ holds. Then there exist multipliers $\lambda_i \geq 0$, $i \in I_0(\bar{x})$, $\mu_j \in \mathbb{R}$, $j \in J$, with*

$$\nabla f(\bar{x}) + \sum_{i \in I_0(\bar{x})} \lambda_i \nabla g_i(\bar{x}) + \sum_{j \in J} \mu_j \nabla h_j(\bar{x}) = 0.$$

Here, the λ_i can be chosen so that $|\{i \in I_0(\bar{x}) | \lambda_i > 0\}| \leq n - q$ holds.

Proof The assertion follows analogously to the proof of Theorem 3.2.40 from Theorem 3.2.22, as soon as we have shown that in the transformed problem $\widetilde{P}$ the MFCQ is satisfied at $\bar{z} = 0$. Indeed, according to part a of the MFCQ in Definition 3.2.41 the problem $\widetilde{P}$ can be constructed in the first place.

Furthermore, according to Definition 3.2.3 the MFCQ at $\bar{z} = 0$ in $\widetilde{P}$ means the existence of some $\delta \in \mathbb{R}^{n-q}$ with

$$\forall i \in I_0(\bar{z}): \quad 0 > \langle \nabla \widetilde{g}_i(\bar{z}), \delta \rangle = \langle \Psi^{\top} \nabla g_i(\bar{x}), \delta \rangle = \langle \nabla g_i(\bar{x}), \Psi \delta \rangle$$

with the matrix Ψ from above, whose columns form an orthonormal basis of the kernel of $Dh(\bar{x})$. Equivalently, one can require the existence of a vector $d \in \ker Dh(\bar{x})$ with

$$\forall i \in I_0(\bar{x}): \quad 0 > \langle \nabla g_i(\bar{x}), d \rangle.$$

In view of the definition of a matrix kernel, this is exactly part b in Definition 3.2.41.   □

In practice, often a regularity condition is used that is even stronger than the MFCQ, because it is easier to check algorithmically.

Definition 3.2.43 (Linear Independence Constraint Qualification) At $\bar{x} \in M$ the *linear independence constraint qualification* (LICQ) holds, if the vectors $\nabla g_i(\bar{x})$, $i \in I_0(\bar{x})$, $\nabla h_j(\bar{x})$, $j \in J$, i.e., the gradients of all active constraints at $\bar{x}$, are linearly independent.

Theorem 3.2.44 *The LICQ at $\bar{x}$ implies the MFCQ at $\bar{x}$.*

Proof At $\bar{x}$, the MFCQ holds if and only if the vectors $\nabla h_j(\bar{x}), j \in J$, are linearly independent and the system $\langle \nabla g_i(\bar{x}), \Psi \delta \rangle < 0, i \in I_0(\bar{x})$, is solvable in δ. According to Gordan's lemma, the latter is equivalent to the equation

$$\sum_{i \in I_0(\bar{x})} \lambda_i \Psi^\top \nabla g_i(\bar{x}) = 0$$

not being solvable with $\lambda_{I_0} \geq 0, \lambda_{I_0} \neq 0$. Overall, the MFCQ can therefore be equivalently written as the simultaneous unsolvability of

$$\sum_{j \in J} \mu_j \nabla h_j(\bar{x}) = 0$$

with $\mu \neq 0$ and

$$\Psi^\top \left(\sum_{i \in I_0(\bar{x})} \lambda_i \nabla g_i(\bar{x}) \right) = 0$$

with $\lambda_{I_0} \geq 0, \lambda_{I_0} \neq 0$. It is not hard to see that this is equivalent to the unsolvability of

$$\sum_{i \in I_0(\bar{x})} \lambda_i \nabla g_i(\bar{x}) + \sum_{j \in J} \mu_j \nabla h_j(\bar{x}) = 0$$

with $\lambda_{I_0} \geq 0$ and $(\lambda_{I_0}, \mu) \neq (0, 0)$. If we drop the condition $\lambda_{I_0} \geq 0$ here, we obtain exactly the LICQ. Consequently, the LICQ is stronger than the MFCQ.   □

Exercise 3.2.45 Construct a feasible set $M \subseteq \mathbb{R}^2$, described by two inequality constraints, at the boundary of which the LICQ is nowhere fulfilled, but the MFCQ holds everywhere.

Example 3.2.46 Theorem 3.2.42 allows, for example, to derive an algorithmically accessible formula for the spectral norm

$$\|A\|_2 = \max\{\|Ad\|_2 \mid \|d\|_2 = 1\}$$

of an (m, n)-matrix A (Remark 2.2.41). This number is the maximal value of the optimization problem

$$P: \quad \max_{d \in \mathbb{R}^n} \|Ad\|_2 \quad \text{s.t.} \quad \|d\|_2 = 1$$

with a continuous objective function and a nonempty and compact feasible set, so according to the Weierstrass theorem, it is attained at some maximal point $\bar{d}$. In view of Exercises 1.1.3 and 1.3.4, the nonsmooth maximization problem P has the same optimal points as the smooth minimization problem

$$Q: \quad \min_{d \in \mathbb{R}^n} -\|Ad\|_2^2 \quad \text{s.t.} \quad \|d\|_2^2 = 1,$$

and the maximal value v_P of P is recovered from the minimal value v_Q of Q as $v_P = \sqrt{-v_Q}$.

The objective function of the problem Q is $f(d) = -d^\mathsf{T} A^\mathsf{T} A d$ with $\nabla f(d) = -2A^\mathsf{T} A d$, and the equality constraint is $0 = h(d) = d^\mathsf{T} d - 1$ with $\nabla h(d) = 2d$. In particular, for every minimal point $\bar{d}$ of Q we have $\nabla h(\bar{d}) \neq 0$, so that the LICQ holds there. Theorem 3.2.42 thus provides for every minimal point $\bar{d}$ of Q the existence of some multiplier $\bar{\mu} \in \mathbb{R}$ with

$$0 = \nabla f(\bar{d}) + \bar{\mu} \nabla h(\bar{d}) = -2A^\mathsf{T} A \bar{d} + 2\bar{\mu} \bar{d}.$$

This is equivalent to $A^\mathsf{T} A \bar{d} = \bar{\mu} \bar{d}$, so that $\bar{d}$ is necessarily an eigenvector of $A^\mathsf{T} A$ with associated eigenvalue $\bar{\mu}$. As the minimal value of Q results

$$v_Q = -\bar{d}^\mathsf{T} A^\mathsf{T} A \bar{d} = -\bar{d}^\mathsf{T} (\bar{\mu} \bar{d}) = -\bar{\mu} \bar{d}^\mathsf{T} \bar{d} = -\bar{\mu}$$

and as the maximal value of P therefore

$$v_P = \sqrt{-v_Q} = \sqrt{\bar{\mu}},$$

where $\bar{\mu}$ coincides with one of the eigenvalues $\lambda_1, \ldots, \lambda_n$ of $A^\mathsf{T} A$.

The application of Theorem 3.2.42 has thus shown that v_P cannot coincide with any value outside of the set $\{\sqrt{\lambda} \mid \lambda$ is eigenvalue of $A^\mathsf{T} A\}$. Among the remaining (at most n) possibilities for v_P it is easy to choose the best one: With the (or some) maximal eigenvalue $\lambda_{\max}(A^\mathsf{T} A)$ of the matrix $A^\mathsf{T} A$ we obtain

$$\|A\|_2 \; = \; v_P \; = \; \sqrt{\lambda_{\max}(A^\mathsf{T} A)}.$$

We had discussed the geometric interpretation of this formula in Remark 2.2.41.

Remark 3.2.47 Every number $\sqrt{\lambda}$ to a positive eigenvalue λ of the matrix $A^\mathsf{T} A$ is called *singular value* of the (m, n)-matrix A.

3.2.8   Karush-Kuhn-Tucker Points

The fundamental Theorem 3.2.42 allows us to transfer the concept of a critical point from unconstrained to constrained problems P.

Definition 3.2.48 (Karush-Kuhn-Tucker Point) For $\bar{x} \in M$ let there exist $\lambda_i \geq 0, i \in I_0(\bar{x}), \mu_j \in \mathbb{R}, j \in J$, with

$$\nabla f(\bar{x}) + \sum_{i \in I_0(\bar{x})} \lambda_i \nabla g_i(\bar{x}) + \sum_{j \in J} \mu_j \nabla h_j(\bar{x}) \; = \; 0.$$

Then $\bar{x}$ is called *Karush-Kuhn-Tucker point (KKT point)* of P. The coefficients $\lambda_i, i \in I_0(\bar{x})$, and $\mu_j, j \in J$, are called *KKT multipliers*.

To prove Theorem 3.2.42, we showed that every stationary point in the sense of Definition 3.2.1, at which the MFCQ holds, is a KKT point. Thus, like the concept of the critical point in unconstrained optimization, the concept of the KKT point plays the role of the 'algebraic translation' of the geometrically motivated and therefore algorithmically hard to verify concept of stationarity.

In the literature, KKT multipliers are often also referred to as *Lagrange multipliers*. Lagrange and Euler already knew precursors of the Karush-Kuhn-Tucker theorem for the case without inequality constraints in the middle of the eighteenth century.

From results of linear algebra it follows that the multipliers of a KKT point $\bar{x}$ are *uniquely* determined if the LICQ holds at $\bar{x}$. Theorems 3.2.42 and 3.2.44 therefore provide the following result.

Corollary 3.2.49 *Let $\bar{x}$ be a local minimal point of P, at which the functions f, g_i, $i \in I_0(\bar{x})$, and h_j, $j \in J$, are continuously differentiable, and at which the LICQ holds. Then $\bar{x}$ is a KKT point of P with unique KKT multipliers.*

With Corollary 3.2.49 we are able to provide, analogous to Algorithm 2.1, the conceptual Algorithm 3.1 for constrained nonlinear minimization. It uses the equivalent reformulation of the statement of Corollary 3.2.49 that at every local minimal point $\bar{x}$ of P one of two cases can occur:

- the LICQ is violated
- the LICQ is fulfilled and at the same time $\bar{x}$ is a KKT point.

Algorithm 3.1: Conceptual algorithm for constrained nonlinear minimization with first-order information

---

    Input: Solvable constrained C^1-optimization problem P
    Output: Global minimal point $x^\star$ of P

1 **begin**
2   Determine the set LD of points in M, at which the LICQ is violated.
3   Determine among the points in M, at which the LICQ holds, the set KKT of all KKT points.
4   Determine a minimal point $x^\star$ of f in $LD \cup KKT$.
5 **end**

Exercise 3.2.50 Use Theorem 3.2.42 to improve Algorithm 3.1. How does the candidate set change if there are elements among the KKT points at which only the MFCQ, but not the LICQ, is satisfied?

To determine a KKT point, one must find a solution (x, λ, μ) of the following system, in which also the feasibility of x is explicitly required:

$$\left.\begin{array}{r}
\nabla f(x) + \sum_{i \in I_0(x)} \lambda_i \nabla g_i(x) + \sum_{j \in J} \mu_j \nabla h_j(x) = 0, \\
g_i(x) \leq 0, \ i \in I, \\
h_j(x) = 0, \ j \in J, \\
\lambda_i \geq 0, \ i \in I_0(x).
\end{array}\right\} \quad (3.11)$$

To avoid algorithmic problems with the domain of the left side of the first equation in (3.11), from now on we assume that the functions f, g and h are continuously differentiable on all of $\mathbb{R}^n$.

Unfortunately, the system (3.11) depends on the a priori unknown active index set $I_0(x)$, so that for the computation of KKT points one must consider case distinctions with respect to $I_0(x)$. Due to $I = \{1, \ldots p\}$, in the worst case there exist 2^p possibilities to choose subsets $I_0(x)$ of I, i.e., *exponentially many*. In view of Carathéodory's theorem, at least it suffices to consider active index sets with $|I_0(x)| \le n - q$. This reduces the number of subcases if $p > n - q$ holds. Still, for practically relevant sizes of p, q and n, the number of subcases can be vast, so that one takes a different algorithmic approach.

For this purpose, one introduces artificial multipliers λ_i for $i \in I \setminus I_0(x)$, and ensures through an additional constraint that these have the value zero. For example, (3.11) is equivalent to

$$\left. \begin{array}{r} \nabla f(x) + \sum_{i \in I} \lambda_i \, \nabla g_i(x) + \sum_{j \in J} \mu_j \, \nabla h_j(x) = 0, \\ g_i(x) \le 0, \ i \in I, \\ h_j(x) = 0, \ j \in J, \\ \lambda_i \ge 0, \ i \in I, \\ \lambda_i \cdot g_i(x) = 0, \ i \in I. \end{array} \right\} \tag{3.12}$$

A condition of the form $a, b \ge 0$, $a \cdot b = 0$ is called a *complementarity condition*. Thus, for all $i \in I$, λ_i and $-g_i(x)$ satisfy a complementarity condition. For every active constraint g_i at x, complementarity is automatically fulfilled, while for every inactive constraint (i.e., for $g_i(x) < 0$) complementarity enforces $\lambda_i = 0$. This makes (3.11) and (3.12) equivalent.

Under the sign constraints on λ_i and $g_i(x)$, the last line of (3.12) could equivalently be replaced by the aggregated condition

$$\sum_{i \in I} \lambda_i \cdot g_i(x) = 0. \tag{3.13}$$

In linear optimization, with the definitions from Example 3.2.10, this leads to the central condition $\lambda^\mathsf{T}(Ax + b) = 0$, where the multiplier vector λ is then called *dual variable* [25, 36]. As we will see shortly, such an aggregation is not desirable in nonlinear optimization.

At a KKT point $\bar{x}$, for each $i \in I$ either $g_i(\bar{x}) = 0$ or $g_i(\bar{x}) < 0$ and either $\lambda_i = 0$ or $\lambda_i > 0$ hold. The complementarity condition enforces $\lambda_i = 0$ for $g_i(\bar{x}) < 0$, but only $\lambda_i \ge 0$ for $g_i(\bar{x}) = 0$, rather than $\lambda_i > 0$. For many results in nonlinear optimization, it is important to tighten this relationship to achieve an equivalence of $\lambda_i = 0$ and $g_i(\bar{x}) < 0$ on the one hand and $\lambda_i > 0$ and $g_i(\bar{x}) = 0$ on the other hand.

Definition 3.2.51 (Strict Complementarity Condition) At the KKT point $\bar{x}$ let the LICQ hold, and let $(\bar{\lambda}_{I_0}, \bar{\mu})$ be the unique solution of (3.11). Then at $\bar{x}$ the *strict complementarity condition (SCC)* holds if $\bar{\lambda}_i > 0$ is fulfilled for all $i \in I_0(\bar{x})$.

Fig. 3.13 Set defined by a
(strict) complementarity
condition

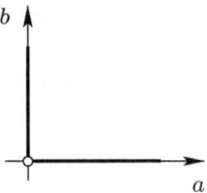

In the optimization literature, this notion is also known as *strict complementary slackness (SCS)*. In the (a, b)-plane, points (a, b) that satisfy a complementarity condition $a, b \geq 0$, $a \cdot b = 0$, form the union of the nonnegative a and b axes (Fig. 3.13). *Strict* complementarity means that the kink point of this set, namely the origin, is removed from the set.

Example 3.2.52 For $f(x) = x^2$ and $g(x) = x$, $\bar{x} = 0$ is a local minimal point of P, at which the LICQ is fulfilled, but the SCC is violated. Indeed, the equation

$$0 = [\, 2x + \lambda \cdot 1\,]_{\bar{x}=0}$$

only allows the solution $\lambda = 0$. We note that the situation at the minimal point is degenerate insofar as f would also have a minimal point there in the absence of the active constraint g. For a more detailed examination of this degeneracy, we refer to [34].

> **Definition 3.2.53 (Lagrange Function)** The function
>
> $$L : \mathbb{R}^n \times \mathbb{R}^p \times \mathbb{R}^q \to \mathbb{R}, \ (x, \lambda, \mu) \mapsto f(x) + \sum_{i \in I} \lambda_i\, g_i(x) + \sum_{j \in J} \mu_j\, h_j(x)$$
>
> is called the *Lagrange function* of P.

With the help of the Lagrange function, one can write (3.12) more compactly as

$$\left. \begin{aligned}
\nabla_x L(x, \lambda, \mu) &= 0, \\
\nabla_\lambda L(x, \lambda, \mu) &\leq 0, \\
\nabla_\mu L(x, \lambda, \mu) &= 0, \\
\lambda &\geq 0, \\
\operatorname{diag}(\lambda) \cdot \nabla_\lambda L(x, \lambda, \mu) &= 0
\end{aligned} \right\} \tag{3.14}$$

where

$$\mathrm{diag}(\lambda) := \begin{pmatrix} \lambda_1 & & 0 \\ & \ddots & \\ 0 & & \lambda_p \end{pmatrix}$$

denotes the diagonal matrix formed with the entries of the vector λ.

In the absence of inequality constraints (i.e., in the case $I = \emptyset$), (3.14) even simplifies to

$$\nabla L(x, \mu) = 0,$$

thus to a critical point condition on the Lagrange function. In duality theory [36], it is shown that under suitable conditions such a critical point of L is a *saddle point* of L. Thus, analogously to the unconstrained case, 'only' an equation needs to be solved. This shows that purely equality-constrained nonlinear optimization problems are algorithmically easier to handle than problems with inequality constraints.

Example 3.2.54 We consider the purely equality-constrained problem

$$P: \qquad \min_{x \in \mathbb{R}^2} f(x) \quad \text{s.t.} \quad h(t, x) = 0$$

with

$$f(x) = \tfrac{1}{2}(x_1^2 - x_2^2),$$
$$h(t, x) = x_2 - \tfrac{t}{2}x_1^2 - 1$$

and a parameter $t \in \mathbb{R}$. The associated Lagrange function is

$$L(x, \mu) = \tfrac{1}{2}(x_1^2 - x_2^2) + \mu \left(x_2 - \tfrac{t}{2}x_1^2 - 1\right),$$

so that the KKT system is

$$0 = \nabla_x L(x, \mu) = \begin{pmatrix} x_1 \\ -x_2 \end{pmatrix} + \mu \begin{pmatrix} -tx_1 \\ 1 \end{pmatrix},$$
$$0 = \nabla_\mu L(x, \mu) = x_2 - \tfrac{t}{2}x_1^2 - 1.$$

Therefore, the point $\bar{x} = (0, 1)^\mathsf{T}$ is a KKT point of P for every $t \in \mathbb{R}$ with $\bar{\mu} = 1$. Because of $\nabla h(\bar{x}) = (0, 1)^\mathsf{T}$, the LICQ also holds at $\bar{x}$ for every $t \in \mathbb{R}$.

Figure 3.14 geometrically shows that the point $\bar{x}$ is actually a local minimal point of P only for sufficiently small t, while it is a local *maximal* point for large values of t.

Fig. 3.14 Level lines of f and feasible sets for two values of t

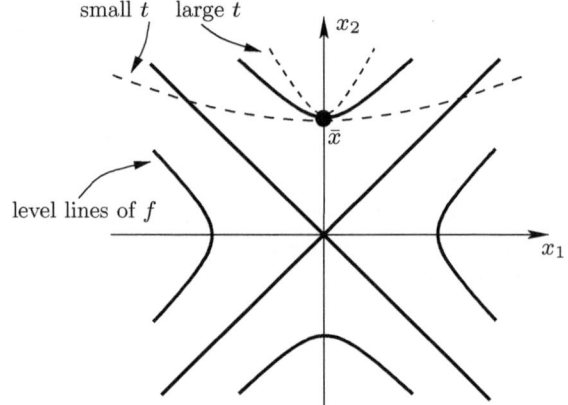

Exercise 3.2.55 In Example 3.2.54, replace the equality constraint $h(t, x) = 0$ with the inequality constraint $h(t, x) \leq 0$ and show that $\bar{x} = (0, 1)^\mathsf{T}$ is a KKT point of the new problem P for every $t \in \mathbb{R}$, but not a local minimal point for too large values of t.

In Example 3.2.54, the critical parameter value $\bar{t}$, at which $\bar{x}$ switches from a local minimal to a maximal point, cannot be determined using the KKT condition as a first-order condition, since the change is due to a critical *curvature* of the constraint function. Fortunately, as in the unconstrained case, such curvature information can be taken into account by second-order conditions.

3.2.9 Second-Order Optimality Conditions

To motivate the form of the second-order optimality conditions in constrained optimization, we consider the equation that results from (3.14) by temporarily omitting the inequalities:

$$0 = \mathscr{T}(x, \lambda, \mu) := \begin{pmatrix} \nabla_x L(x, \lambda, \mu) \\ \operatorname{diag}(\lambda) \cdot \nabla_\lambda L(x, \lambda, \mu) \\ \nabla_\mu L(x, \lambda, \mu) \end{pmatrix}. \tag{3.15}$$

Definition 3.2.56 (Critical Point of a Constrained Problem) A point $\bar{x} \in M$, which fulfills the equation $\mathscr{T}(\bar{x}, \bar{\lambda}, \bar{\mu}) = 0$ with multipliers $\bar{\lambda}$ and $\bar{\mu}$, is called a *critical point* of P.

Example 3.2.57 In Example 3.2.54, the point $\bar{x} = (0, 1)^\top$ is a critical point of P for every $t \in \mathbb{R}$.

Note that zeros $(\bar{x}, \bar{\lambda}, \bar{\mu})$ of $\mathscr{T}$ can be infeasible in contrast to critical points, and that critical points only differ from KKT points in that no nonnegativity condition is required for $\bar{\lambda}$. In (3.15), there are exactly $n + p + q$ equations for $n + p + q$ unknowns, so that one can expect isolated zeros of $\mathscr{T}$ under a suitable regularity condition. If the complementarity conditions had been aggregated as in (3.13), $p - 1$ equations would be missing for this argument.

Crucial for the solution structure of (3.14) is whether the Jacobian matrix $D\mathscr{T}(x, \lambda, \mu)$ is nonsingular at a zero of $\mathscr{T}$. We therefore assume all occurring functions to be twice continuously differentiable from now on and calculate this Jacobian matrix. To this end, without loss of generality, let $I_0(\bar{x}) = \{1, \ldots, p_0\}$, and we set

$$g_{I_0} := \begin{pmatrix} g_1 \\ \vdots \\ g_{p_0} \end{pmatrix}, \qquad \lambda_{I_0} := \begin{pmatrix} \lambda_1 \\ \vdots \\ \lambda_{p_0} \end{pmatrix},$$

$$g_{I_0^c} := \begin{pmatrix} g_{p_0+1} \\ \vdots \\ g_p \end{pmatrix}, \qquad \lambda_{I_0^c} := \begin{pmatrix} \lambda_{p_0+1} \\ \vdots \\ \lambda_p \end{pmatrix}.$$

In the function $\mathscr{T}$, we can thus distinguish between active and inactive constraints by

$$\mathscr{T}(x, \lambda, \mu) = \begin{pmatrix} \nabla_x L(x, \lambda, \mu) \\ \mathrm{diag}(\lambda_{I_0}) \cdot g_{I_0}(x) \\ \mathrm{diag}(\lambda_{I_0^c}) \cdot g_{I_0^c}(x) \\ h(x) \end{pmatrix}.$$

Lemma 3.2.58 *Let* $\mathscr{T}(\bar{x}, \bar{\lambda}, \bar{\mu}) = 0$. *Then* $D\mathscr{T}(\bar{x}, \bar{\lambda}, \bar{\mu})$ *is nonsingular if and only if* $\bar{\lambda}_i \neq 0$ *holds for all* $i \in I_0(\bar{x})$ *and if the matrix*

$$\begin{pmatrix} D_x^2 L(\bar{x}, \bar{\lambda}, \bar{\mu}) & \nabla g_{I_0}(\bar{x}) & \nabla h(\bar{x}) \\ D g_{I_0}(\bar{x}) & 0 & 0 \\ D h(\bar{x}) & 0 & 0 \end{pmatrix}$$

is nonsingular.

Proof In the following proof, we omit for abbreviation the obvious arguments of the occurring functions. It holds

$$D\mathcal{T}(\bar{x}, \bar{\lambda}, \bar{\mu}) = D_{(x, \lambda_{I_0}, \lambda_{I_0^c}, \mu)}\mathcal{T}(\bar{x}, \bar{\lambda}_{I_0}, \bar{\lambda}_{I_0^c}, \bar{\mu})$$

$$= \begin{pmatrix} D_x^2 L & \nabla g_{I_0} & \nabla g_{I_0^c} & \nabla h \\ \mathrm{diag}(\bar{\lambda}_{I_0})Dg_{I_0} & \mathrm{diag}(g_{I_0}) & 0 & 0 \\ \mathrm{diag}(\bar{\lambda}_{I_0^c})Dg_{I_0^c} & 0 & \mathrm{diag}(g_{I_0^c}) & 0 \\ Dh & 0 & 0 & 0 \end{pmatrix}.$$

The block $\mathrm{diag}(g_{I_0})$ of this matrix disappears due to the definition of an active index. The matrix $\mathrm{diag}(\bar{\lambda}_{I_0^c})$ also disappears, since due to $\mathcal{T} = 0$ the complementarity conditions are fulfilled in particular. Swapping the third and fourth block rows and the third and fourth block columns in the resulting matrix shows that $D\mathcal{T}$ is nonsingular if and only if the matrix

$$\left(\begin{array}{ccc|c} D_x^2 L & \nabla g_{I_0} & \nabla h & \nabla g_{I_0^c} \\ \mathrm{diag}(\bar{\lambda}_{I_0})Dg_{I_0} & 0 & 0 & 0 \\ Dh & 0 & 0 & 0 \\ \hline 0 & 0 & 0 & \mathrm{diag}(g_{I_0^c}) \end{array} \right)$$

is. From the definition of an inactive index it follows that none of the entries on the diagonal of $\mathrm{diag}(g_{I_0^c})$ disappear, so this diagonal matrix is nonsingular. The block structure of the above matrix therefore implies that it is nonsingular if and only if

$$\begin{pmatrix} D_x^2 L & \nabla g_{I_0} & \nabla h \\ \mathrm{diag}(\bar{\lambda}_{I_0})Dg_{I_0} & 0 & 0 \\ Dh & 0 & 0 \end{pmatrix}$$

is. The latter matrix can be written as the product of

$$\begin{pmatrix} I_n & 0 & 0 \\ 0 & \mathrm{diag}(\bar{\lambda}_{I_0}) & 0 \\ 0 & 0 & I_q \end{pmatrix}$$

and

$$\begin{pmatrix} D_x^2 L & \nabla g_{I_0} & \nabla h \\ Dg_{I_0} & 0 & 0 \\ Dh & 0 & 0 \end{pmatrix},$$

so that finally $D\mathcal{T}$ is nonsingular if and only if these two matrices are nonsingular. This proves the claim.   □

Next, we characterize the nonsingularity of matrices of the structure

$$\begin{pmatrix} A & B \\ B^\mathsf{T} & 0 \end{pmatrix},$$

where A is an (n, n)-matrix and B is an (n, m)-matrix, for example $A = D_x^2 L(\bar{x}, \bar{\lambda}, \bar{\mu})$ and $B = (\nabla g_{I_0}(\bar{x}), \nabla h(\bar{x}))$ with $m = p_0 + q$.

For $\operatorname{rank}(B) = m$, $\ker B^\mathsf{T}$ is known to have the dimension $n - m$. First, let $m < n$. Then we choose vectors $v^1, \ldots, v^{n-m} \in \mathbb{R}^n$ which form a basis of $\ker B^\mathsf{T}$, and combine them to the $(n, n-m)$-matrix $V = (v^1, \ldots, v^{n-m})$.

> **Definition 3.2.59 (Restriction of a Matrix)** The $(n-m, n-m)$-matrix $A|_{\ker B^\mathsf{T}} := V^\mathsf{T} A V$ is called *restriction* of A to $\ker B^\mathsf{T}$.

The terminology introduced in Definition 3.2.59 is explained by the fact that the restriction $q|_{\ker B^\mathsf{T}}$ of the quadratic function $q(d) = d^\mathsf{T} A d$ to the space $\ker B^\mathsf{T}$ has the representation $q|_{\ker B^\mathsf{T}}(d) = d^\mathsf{T}(A|_{\ker B^\mathsf{T}})d$.

In the following, we will only be interested in properties of $A|_{\ker B^\mathsf{T}}$ that do not depend on the choice of V. According to Sylvester's law of inertia [8, 20], this includes, for example, the numbers of positive, negative, and vanishing eigenvalues of $A|_{\ker B^\mathsf{T}}$.

In the case $m = n$, $\operatorname{rank}(B) = m$ is equivalent to the nonsingularity of B. Then $\ker B^\mathsf{T} = \{0\}$ holds, and $A|_{\ker B^\mathsf{T}}$ formally becomes a '$(0, 0)$-matrix' (i.e., a matrix without any rows and columns). Formally, such an empty matrix is not only nonsingular but even positive as well as negative definite, because the corresponding statements about all eigenvalues are trivially fulfilled. This formal view will also turn out to be meaningful in optimality conditions.

Example 3.2.60 For $B = (\nabla g_{I_0}(\bar{x}), \nabla h(\bar{x}))$, $\operatorname{rank}(B) = m = p_0 + q$ holds if and only if the LICQ at $\bar{x}$ is fulfilled. In this case,

$$\ker B^\mathsf{T} = \{d \in \mathbb{R}^n | \langle \nabla g_i(\bar{x}), d \rangle = 0, \ i \in I_0(\bar{x}), \ \langle \nabla h_j(\bar{x}), d \rangle = 0, \ j \in J\}$$

$$=: T(\bar{x}, M)$$

is called *tangent space* to M at $\bar{x}$. For $A = D_x^2 L(\bar{x}, \bar{\lambda}, \bar{\mu})$,

$$A|_{\ker B^\mathsf{T}} = D_x^2 L(\bar{x}, \bar{\lambda}, \bar{\mu})|_{T(\bar{x}, M)}$$

is the Hessian matrix of the Lagrange function restricted to the tangent space to M at $\bar{x}$. In the case $p_0 + q = n$ it formally is a $(0, 0)$-matrix.

Lemma 3.2.61 (Structural Lemma)   *For an $(n,\ n)$-matrix A and an $(n,\ m)$-matrix B, the block matrix*

$$\begin{pmatrix} A & B \\ B^\mathsf{T} & 0 \end{pmatrix}$$

is nonsingular if and only if $\mathrm{rank}(B) = m$ *holds and* $A|_{\ker B^\mathsf{T}}$ *is nonsingular.*

Proof First, let

$$\begin{pmatrix} A & B \\ B^\mathsf{T} & 0 \end{pmatrix}$$

be nonsingular. Then, in particular, the last m columns of this matrix are linearly independent, so $\mathrm{rank}(B) = m$ holds. This necessarily also implies $m \le n$. In the case $m = n$, $A|_{\ker B^\mathsf{T}}$ is formally a $(0,0)$-matrix, so it is trivially nonsingular.

Next, let $m < n$ and V be a $(n, n-m)$-matrix, whose columns form a basis of $\ker B^\mathsf{T}$. Then $A|_{\ker B^\mathsf{T}} = V^\mathsf{T} A V$ holds, and we have to show that $V^\mathsf{T} A V$ is nonsingular. For this we choose some $\eta \in \mathbb{R}^{n-m} \setminus \{0\}$. We need to show $V^\mathsf{T} A V \eta \neq 0$.

We set $d := V\eta$. Since V has full column rank, $d \neq 0$ also holds. Suppose that $V^\mathsf{T} A V \eta = 0$ holds. This means

$$A V \eta \in \ker V^\mathsf{T} = (\mathrm{img}\,V)^\perp = \mathrm{img}\,B,$$

so there exists some $\theta \in \mathbb{R}^m$ with $A V \eta = B\theta$. Thus we obtain

$$\begin{pmatrix} A & B \\ B^\mathsf{T} & 0 \end{pmatrix} \begin{pmatrix} d \\ -\theta \end{pmatrix} = \begin{pmatrix} A V \eta - B\theta \\ B^\mathsf{T} V \eta \end{pmatrix} = \begin{pmatrix} 0 \\ 0 \end{pmatrix},$$

and by assumption all entries of d and $-\theta$ must therefore vanish. This contradicts $d \neq 0$, and $V^\mathsf{T} A V$ is therefore nonsingular.

To prove the converse, let $\mathrm{rank}(B) = m$ and $A|_{\ker B^\mathsf{T}}$ be nonsingular. The rank condition on B implies $m \le n$. In the case $m = n$, B is square, so the nonsingularity of the matrix

$$\begin{pmatrix} A & B \\ B^\mathsf{T} & 0 \end{pmatrix}$$

follows from its block structure. Next, let $m < n$. We choose vectors $d \in \mathbb{R}^n$ and $c \in \mathbb{R}^m$ with $(d, c) \neq 0$ and need to show

$$\begin{pmatrix} A & B \\ B^\mathsf{T} & 0 \end{pmatrix} \begin{pmatrix} d \\ c \end{pmatrix} \neq 0.$$

Suppose the latter is not the case. Then

$$\begin{pmatrix} 0 \\ 0 \end{pmatrix} = \begin{pmatrix} A & B \\ B^\mathsf{T} & 0 \end{pmatrix} \begin{pmatrix} d \\ c \end{pmatrix} = \begin{pmatrix} Ad + Bc \\ B^\mathsf{T} d \end{pmatrix} \tag{3.16}$$

holds. From the second of the two equations in (3.16) we deduce $d \in \ker B^\mathsf{T}$, so there exists some $\eta \in \mathbb{R}^{n-m}$ with $d = V\eta$. Therefore, the first equation gives

$$0 = V^\mathsf{T} \underbrace{(Ad + Bc)}_{=0} = V^\mathsf{T} A V\eta + \underbrace{V^\mathsf{T} B c}_{=0} = V^\mathsf{T} A V\eta.$$

The nonsingularity of $V^\mathsf{T} A V$ implies $\eta = 0$ and thus $d = V\eta = 0$.

The first equation of (3.16) thus reduces to $0 = Bc$. Because of $\mathrm{rank}(B) = m$, this means $c = 0$, and overall $(d, c) = 0$. This contradicts the choice of (d, c), and the claim is proven. $\qquad\square$

Theorem 3.2.62 *Let $\mathscr{T}(\bar{x}, \bar{\lambda}, \bar{\mu}) = 0$. Then $D\mathscr{T}(\bar{x}, \bar{\lambda}, \bar{\mu})$ is nonsingular if and only if the following three conditions are simultaneously met:*

(a) *The LICQ holds at $\bar{x}$.*
(b) *$\bar{\lambda}_i \neq 0$ holds for all $i \in I_0(\bar{x})$.*
(c) *The matrix $D_x^2 L(\bar{x}, \bar{\lambda}, \bar{\mu})|_{T(\bar{x}, M)}$ is nonsingular.*

Proof Lemmas 3.2.58 and 3.2.61. $\qquad\square$

In the case $n = p_0 + q$, condition c in Theorem 3.2.62 is trivially fulfilled and can be omitted.

Definition 3.2.63 (Nondegenerate Critical Point of a Constrained Problem) For a critical point $\bar{x}$ of P with multipliers $\bar{\lambda}, \bar{\mu}$, let conditions a, b, and c from Theorem 3.2.62 be fulfilled. Then $\bar{x}$ is called a *nondegenerate critical point* of P.

Example 3.2.64 In Example 3.2.54, the critical point $\bar{x} = (0, 1)^\mathsf{T}$ of P fulfills the LICQ, so that statement a from Theorem 3.2.62 holds. Statement b does not need to be considered, as there are no inequality constraints. To check condition c, we calculate

$$D^2 f(x) = \begin{pmatrix} 1 & 0 \\ 0 & -1 \end{pmatrix} \quad \text{and} \quad D^2 h(x) = \begin{pmatrix} -t & 0 \\ 0 & 0 \end{pmatrix},$$

thus

$$D_x^2 L(\bar{x}, \bar{\mu}) = \begin{pmatrix} 1 & 0 \\ 0 & -1 \end{pmatrix} + 1 \cdot \begin{pmatrix} -t & 0 \\ 0 & 0 \end{pmatrix} = \begin{pmatrix} 1 - t & 0 \\ 0 & -1 \end{pmatrix}.$$

A basis of the tangent space

$$T(\bar{x}, M) = \left\{ d \in \mathbb{R}^2 \middle| 0 = \langle \nabla h(\bar{x}), d \rangle = d_2 \right\} = \mathbb{R} \times \{0\}$$

is given by $v^1 = (1, 0)^\mathsf{T}$, and with $V = v^1$ we obtain

$$D_x^2 L(\bar{x}, \bar{\mu})|_{T(\bar{x},M)} = \begin{pmatrix} 1 \\ 0 \end{pmatrix}^\mathsf{T} \begin{pmatrix} 1-t & 0 \\ 0 & -1 \end{pmatrix} \begin{pmatrix} 1 \\ 0 \end{pmatrix} = 1 - t.$$

Therefore, $D_x^2 L(\bar{x}, \bar{\mu})|_{T(\bar{x},M)}$ is nonsingular for all $t \neq 1$, and $\bar{x}$ is a nondegenerate critical point of P for all $t \neq 1$.

Like in the unconstrained case (Sect. 2.1.4), it can be shown that for 'almost all' C^2-optimization problems P every critical point is nondegenerate. This means that the conditions a, b and c from Theorem 3.2.62 are *mild* (for details see [22]). Since in Definition 3.2.63 neither the multipliers $\bar{\lambda}_i$, $i \in I_0(\bar{x})$, nor the eigenvalues of the matrix $D_x^2 L(\bar{x}, \bar{\lambda}, \bar{\mu})|_{T(\bar{x},M)}$ are required to be positive, nondegenerate critical points can be not only minimal points, but also maximal points or saddle points.

To proceed to minimal points, we need to reintroduce these sign constraints.

Definition 3.2.65 (Nondegenerate Minimal Point of a Constrained Problem) Let a KKT point $\bar{x}$ of P with multipliers $\bar{\lambda}, \bar{\mu}$ satisfy the following conditions:

(a) At $\bar{x}$ the LICQ holds.
(b) At $\bar{x}$ the SCC holds.
(c) The matrix $D_x^2 L(\bar{x}, \bar{\lambda}, \bar{\mu})|_{T(\bar{x},M)}$ is positive definite.

Then $\bar{x}$ is called a *nondegenerate local minimal point* of P.

Condition c in Definition 3.2.65 is equivalent to the statement

$$\forall d \in T(\bar{x}, M) \setminus \{0\}: \quad d^\mathsf{T} D_x^2 L(\bar{x}, \bar{\lambda}, \bar{\mu}) d > 0.$$

Again, in the case $n = p_0 + q$ it is trivially fulfilled and can be omitted.

Example 3.2.66 In Example 3.2.54, due to

$$D_x^2 L(\bar{x}, \bar{\mu})|_{T(\bar{x},M)} = 1 - t,$$

the critical point $\bar{x} = (0, 1)^\mathsf{T}$ of P is a nondegenerate local minimal point of P for all $t < 1$.

The following result confirms that nondegenerate local minimal points are indeed local minimal points.

> **Theorem 3.2.67 (Second-Order Sufficient Optimality Condition)** *Every nondegenerate local minimal point is a strict local minimal point of P.*

Proof Let $\bar{x}$ be a nondegenerate local minimal point with multipliers $\bar{\lambda}, \bar{\mu}$. Suppose $\bar{x}$ is not a strict local minimal point of P. Then there exists a sequence $x^k \to \bar{x}$ with $x^k \in M \setminus \{\bar{x}\}$ and $f(x^k) \le f(\bar{x})$ for all $k \in \mathbb{N}$. Thus, without loss of generality, there are sequences $t^k \searrow 0$ and $d^k \to \bar{d}$ with $\|\bar{d}\| = 1$ and $\bar{x} + t^k d^k \in M$ as well as $f(\bar{x} + t^k d^k) \le f(\bar{x})$ for all $k \in \mathbb{N}$. Taylor's theorem provides

$$
\left.
\begin{aligned}
0 &\ge f(\bar{x} + t^k d^k) - f(\bar{x}) \\
&= t^k \langle \nabla f(\bar{x}), d^k \rangle + \frac{(t^k)^2}{2}(d^k)^\mathsf{T} D^2 f(\bar{x}) d^k + o((t^k)^2), \\
\forall\, i \in I_0(\bar{x}): \quad 0 &\ge g_i(\bar{x} + t^k d^k) - \underbrace{g_i(\bar{x})}_{=\,0} \\
&= t^k \langle \nabla g_i(\bar{x}), d^k \rangle + \frac{(t^k)^2}{2}(d^k)^\mathsf{T} D^2 g_i(\bar{x}) d^k + o((t^k)^2), \\
\forall\, j \in J: \quad 0 &= h_j(\bar{x} + t^k d^k) - h_j(\bar{x}) \\
&= t^k \langle \nabla h_j(\bar{x}), d^k \rangle + \frac{(t^k)^2}{2}(d^k)^\mathsf{T} D^2 h_j(\bar{x}) d^k + o((t^k)^2).
\end{aligned}
\right\}
\tag{3.17}
$$

In the following we will use that, as a nondegenerate local minimal point, $\bar{x}$ is a KKT point of P at which the LICQ and the SCC hold. We will create the required contradiction by showing that then $D_x^2 L(\bar{x}, \bar{\lambda}, \bar{\mu})|_{T(\bar{x},M)}$ cannot be positive definite.

Multiplication of the inequalities and equalities in (3.17) with the corresponding multipliers $\bar{\lambda}_i$ and $\bar{\mu}_j$ and subsequent summation results in

$$
0 \ge t^k \underbrace{\left\langle \nabla f(\bar{x}) + \sum_{i \in I_0(\bar{x})} \bar{\lambda}_i \nabla g_i(\bar{x}) + \sum_{j \in J} \bar{\mu}_j \nabla h_j(\bar{x}),\, d^k \right\rangle}_{=\,0}
$$

$$
+ \frac{(t^k)^2}{2}(d^k)^\mathsf{T} \left(D^2 f(\bar{x}) + \sum_{i \in I_0(\bar{x})} \bar{\lambda}_i D^2 g_i(\bar{x}) + \sum_{j \in J} \bar{\mu}_j D^2 h_j(\bar{x}) \right) d^k
$$

$$
+ o((t^k)^2).
$$

Division by $(t^k)^2/2$ and taking the limit $k \to \infty$ yield

$$0 \geq \bar{d}^\mathsf{T} D_x^2 L(\bar{x}, \bar{\lambda}, \bar{\mu}) \bar{d}, \tag{3.18}$$

where the inactive inequalities are taken into account by multipliers $\bar{\lambda}_i = 0$, $i \in I \setminus I_0(\bar{x})$. So far, we have shown that the matrix $D_x^2 L(\bar{x}, \bar{\lambda}, \bar{\mu})$ is not positive definite. However, this does not yet exclude that it is at least positive definite on the subspace $T(\bar{x}, M)$.

To also exclude this, we draw a further conclusion from (3.17) by dividing the individual inequalities by t^k and subsequently taking the limit $k \to \infty$. This yields the relations

$$0 \geq \langle \nabla f(\bar{x}), \bar{d} \rangle, \quad 0 \geq \langle \nabla g_i(\bar{x}), \bar{d} \rangle, \ i \in I_0(\bar{x}), \quad 0 = \langle \nabla h_j(\bar{x}), \bar{d} \rangle, \ j \in J. \tag{3.19}$$

Because of

$$0 = \left\langle \nabla f(\bar{x}) + \underbrace{\sum_{i \in I_0(\bar{x})} \bar{\lambda}_i \nabla g_i(\bar{x}) + \sum_{j \in J} \bar{\mu}_j \nabla h_j(\bar{x})}_{= 0}, \ \bar{d} \right\rangle$$

$$= \underbrace{\langle \nabla f(\bar{x}), \bar{d} \rangle}_{\leq 0} + \sum_{i \in I_0(\bar{x})} \bar{\lambda}_i \underbrace{\langle \nabla g_i(\bar{x}), \bar{d} \rangle}_{\leq 0} + \sum_{j \in J} \bar{\mu}_j \underbrace{\langle \nabla h_j(\bar{x}), \bar{d} \rangle}_{= 0}$$

it even follows

$$0 = \langle \nabla f(\bar{x}), \bar{d} \rangle, \quad 0 = \bar{\lambda}_i \cdot \langle \nabla g_i(\bar{x}), \bar{d} \rangle, \ i \in I_0(\bar{x}), \quad 0 = \langle \nabla h_j(\bar{x}), \bar{d} \rangle, \ j \in J. \tag{3.20}$$

The SCC implies $0 = \langle \nabla g_i(\bar{x}), \bar{d} \rangle$, $i \in I_0(\bar{x})$. Together with $\|\bar{d}\| = 1$ we therefore obtain $\bar{d} \in T(\bar{x}, M) \setminus \{0\}$. With regard to the subsequent Corollary 3.2.68, we note that in the present case the first equation in (3.20) is redundant due to $\nabla_x L(\bar{x}, \bar{\lambda}, \bar{\mu}) = 0$, thus it does not carry any information that can be used in the following.

Case 1: $p_0 + q = n$. Since under the LICQ $T(\bar{x}, M) = \{0\}$ holds, a contradiction arises to the construction of the vector $\bar{d} \in T(\bar{x}, M) \setminus \{0\}$.

Case 2: $p_0 + q < n$. For $\bar{d} \in T(\bar{x}, M) \setminus \{0\}$, (3.18) holds, i.e., the restriction of $D_x^2 L(\bar{x}, \bar{\lambda}, \bar{\mu})$ to $T(\bar{x}, M)$ is not positive definite, as desired for the contradiction. Formally, this can be seen as follows: Let V be a matrix whose columns form a basis of $T(\bar{x}, M)$. Then $D_x^2 L(\bar{x}, \bar{\lambda}, \bar{\mu})|_{T(\bar{x},M)} = V^\intercal D_x^2 L(\bar{x}, \bar{\lambda}, \bar{\mu})V$ holds, and for $\bar{d}$ there exists some $\eta \in \mathbb{R}^{n-p_0-q}$ with $\bar{d} = V\eta$. Since the columns of V are linearly independent, with $\bar{d} \neq 0$ also η cannot be zero. Thus, by (3.18) $\eta \neq 0$ satisfies

$$\eta^\intercal V^\intercal D_x^2 L(\bar{x}, \bar{\lambda}, \bar{\mu})V\eta \leq 0,$$

and the matrix $D_x^2 L(\bar{x}, \bar{\lambda}, \bar{\mu})|_{T(\bar{x},M)}$ is not positive definite. $\qquad\square$

For the following second-order optimality condition, which does *not* require the assumptions of the LICQ and SCC, let

$$I_{0+}(\bar{x}) := \{i \in I_0(\bar{x}) \mid \bar{\lambda}_i > 0\},$$

$$I_{00}(\bar{x}) := \{i \in I_0(\bar{x}) \mid \bar{\lambda}_i = 0\}$$

and

$$
\begin{aligned}
K(\bar{x}, M) = \{d \in \mathbb{R}^n \mid\ & \langle \nabla f(\bar{x}), d \rangle = 0, \\
& \langle \nabla g_i(\bar{x}), d \rangle = 0,\ i \in I_{0+}(\bar{x}), \\
& \langle \nabla g_i(\bar{x}), d \rangle \leq 0,\ i \in I_{00}(\bar{x}), \\
& \langle \nabla h_j(\bar{x}), d \rangle = 0,\ j \in J\}.
\end{aligned}
$$

Corollary 3.2.68 *Let the KKT point $\bar{x}$ of P with multipliers $\bar{\lambda}, \bar{\mu}$ satisfy $d^\intercal D_x^2 L(\bar{x}, \bar{\lambda}, \bar{\mu})d > 0$ for all $d \in K(\bar{x}, M) \setminus \{0\}$. Then $\bar{x}$ is a strict local minimal point of P.*

Proof As in the proof by contradiction to Theorem 3.2.67, one obtains the existence of a vector $\bar{d}$ with (3.18), (3.19) and (3.20). For $i \in I_{00}(\bar{x})$, however, (3.20) does not provide any new information about $\langle \nabla g_i(\bar{x}), \bar{d} \rangle$, so one must stick to the inequality $\langle \nabla g_i(\bar{x}), \bar{d} \rangle \leq 0$ from (3.19). In this case, on the other hand, the first equation in (3.20) is not necessarily redundant, so that $\langle \nabla f(\bar{x}), d \rangle = 0$ can be included in the definition of the cone $K(\bar{x}, M)$. For the vector $\bar{d}$, therefore, according to (3.18) on the one hand $0 \geq \bar{d}^\intercal D_x^2 L(\bar{x}, \bar{\lambda}, \bar{\mu})\bar{d}$ and on the other hand $\bar{d} \in K(\bar{x}, M) \setminus \{0\}$ hold. This contradicts the assumption. $\qquad\square$

Observe that the missing assumption of the SCC in Corollary 3.2.68 is compensated by the fact that $D_x^2 L(\bar{x}, \bar{\lambda}, \bar{\mu})$ must be positive definite on a possibly larger set than in Theorem 3.2.67.

Corollary 3.2.69 (First-Order Sufficient Optimality Condition) *Let the KKT point $\bar{x}$ of P satisfy the following conditions:*

(a) *At $\bar{x}$ the LICQ holds.*
(b) *At $\bar{x}$ the SCC holds.*
(c) *The identity $p_0 + q = n$ holds.*

Then $\bar{x}$ is a strict local minimal point of P.

Proof Under condition c, $D_x^2 L(\bar{x}, \bar{\lambda}, \bar{\mu})$ is trivially positive definite on $T(\bar{x}, M)$, so $\bar{x}$ is a nondegenerate critical point. The assertion thus follows from Theorem 3.2.67. Alternatively, one can conduct an explicit proof by following the proof of Theorem 3.2.67 up to the case distinction. The contradiction from the first case completes the proof of this corollary.   □

For completeness, we also provide a sufficient first-order optimality condition that does not use constraint qualifications or Karush-Kuhn-Tucker points, but only Fritz-John points (Theorem 3.2.40) with a certain range property. The involved functions only need to be differentiable. This useful result is pointed out in [16, Th. 9.7].

Theorem 3.2.70 *For $\bar{x} \in M$ let there exist multipliers $\bar{\kappa} \geq 0, \bar{\lambda}_i \geq 0, i \in I_0(\bar{x}), \bar{\mu}_j \in \mathbb{R}, j \in J$, with*

$$\bar{\kappa} \nabla f(\bar{x}) + \sum_{i \in I_0(\bar{x})} \bar{\lambda}_i \nabla g_i(\bar{x}) + \sum_{j \in J} \bar{\mu}_j \nabla h_j(\bar{x}) = 0$$

and such that the vectors

$$\bar{\kappa} \nabla f(\bar{x},), \quad \bar{\lambda}_i \nabla g_i(\bar{x}), \ i \in I_0(\bar{x}), \quad \bar{\mu}_j \nabla h_j(\bar{x}), \ j \in J,$$

together possess rank n. Then $\bar{x}$ is a strict local minimal point of P.

Proof As in the proof by contradiction to Theorem 3.2.67, we assume that $\bar{x}$ is not a strict local minimal point of P. As there, from first-order Taylor expansions of the involved functions, the existence of a vector $\bar{d} \neq 0$ with (3.19) follows. With the help of the Fritz-John condition, the equations

$$0 = \langle \bar{\kappa} \nabla f(\bar{x}), \bar{d} \rangle, \quad 0 = \langle \bar{\lambda}_i \nabla g_i(\bar{x}), \bar{d} \rangle, \ i \in I_0(\bar{x}), \quad 0 = \langle \bar{\mu}_j \nabla h_j(\bar{x}), \bar{d} \rangle, \ j \in J.$$

follow from this, analogous to (3.20). Due to the assumed rank condition on the involved vectors, the contradiction $\bar{d} = 0$ follows from this.   □

Recall from Remark 2.1.34 that sufficient first-order optimality conditions for *unconstrained* smooth optimization problems cannot exist. As we have just seen, however, they make sense in the constrained case.

In addition to the *sufficient* optimality conditions, there is also a *necessary* second-order optimality condition, analogous to the unconstrained case. As in the unconstrained case, necessary and sufficient conditions differ by the strictness of the occurring inequalities.

Theorem 3.2.71 (Second-Order Necessary Optimality Condition) *Let $\bar{x}$ be a local minimal point of P, at which the LICQ holds. Then:*

(a) *$\bar{x}$ is a KKT point of P with unique multipliers $\bar{\lambda} \geq 0$ and $\bar{\mu}$.*
(b) *The matrix $D_x^2 L(\bar{x}, \bar{\lambda}, \bar{\mu})|_{T(\bar{x}, M)}$ is positive semidefinite.*

Proof Part a is just the statement of Corollary 3.2.49. The basic idea for proving part b is to construct, for some arbitrary $d \in T(\bar{x}, M)$, a smooth curve $\{x(t)|\ t \in [0, \check{t})\} \subseteq M$ with $x(0) = \bar{x}$ and $\dot{x}(0) = d$, and to exploit the minimality of $\bar{x}$ along this curve.

Since M in general possesses a complicated nonlinear structure, we construct the desired curve (similar to the proof to Theorem 3.2.40) in new coordinates. Because of the validity of LICQ there exist $\eta_{p_0+q+1}, \ldots, \eta_n \in \mathbb{R}^n$, such that the vectors

$$\nabla g_i(\bar{x}),\ i \in I_0(\bar{x}),\ \nabla h_j(\bar{x}),\ j \in J,\ \eta_{p_0+q+1}, \ldots, \eta_n$$

form a basis of $\mathbb{R}^n$. According to the inverse function theorem, then

$$\Phi(x) = \begin{pmatrix} g_{I_0}(x) \\ h(x) \\ \eta_{p_0+q+1}^\mathsf{T}(x - \bar{x}) \\ \vdots \\ \eta_n^\mathsf{T}(x - \bar{x}) \end{pmatrix}$$

is a C^2-diffeomorphism between a neighborhood U of $\bar{x}$ and a neighborhood V of $0 \in \mathbb{R}^n$. Theorem 3.1.3 guarantees for sufficiently small U

$$x \in M \cap U \Leftrightarrow x \in U \cap \{x \in \mathbb{R}^n|\ g_{I_0}(x) \leq 0,\ h(x) = 0\}$$

$$\Leftrightarrow x = \Phi^{-1}(y)$$

$$\text{with}\quad y \in V,\ y_1, \ldots, y_{p_0} \leq 0,\ y_{p_0+1} = \ldots = y_q = 0.$$

Locally around $\bar{x}$, M is therefore C^2-diffeomorphic to the set

$$\widetilde{M} = \{y \in \mathbb{R}^n|\ y_1, \ldots, y_{p_0} \leq 0,\ y_{p_0+1} = \ldots = y_q = 0\},$$

where the points $\bar{x} \in M$ and $0 \in \widetilde{M}$ correspond to each other.

The simple structure of $\widetilde{M}$ allows us to construct a curve contained in $\widetilde{M}$. To this end, let $d \in T(\bar{x}, M)$ be arbitrary. We set

$$\widetilde{y} := D\Phi(\bar{x}) \cdot d = \begin{pmatrix} Dg_{I_0}(\bar{x})d \\ Dh(\bar{x})d \\ \eta_{p_0+q+1}^{\mathsf{T}}d \\ \vdots \\ \eta_n^{\mathsf{T}}d \end{pmatrix} = \begin{pmatrix} 0_p \\ 0_q \\ \star \\ \vdots \\ \star \end{pmatrix}$$

and note that $\widetilde{y}$ lies in $\widetilde{M}$. Since $\widetilde{M}$ is a cone containing the origin, for all $t \geq 0$, $y(t) = t\widetilde{y}$ also lies in $\widetilde{M}$. The ray $\{y(t) \mid t \geq 0\} \subseteq \widetilde{M}$ is the desired curve in new coordinates. To transform it back to the original coordinates, we define for those t with $y(t) \in V$, i.e., for $t \in [0, \check{t})$ with some $\check{t} > 0$,

$$x(t) := \Phi^{-1}(y(t)) = \Phi^{-1}(t\widetilde{y}).$$

Indeed, it holds

$$x(0) = \Phi^{-1}(0) = \bar{x},$$

$$\dot{x}(0) = D[\Phi^{-1}(0)] \cdot \widetilde{y} = (D\Phi(\bar{x}))^{-1}\widetilde{y} = d,$$

and $x(t)$ lies in M for all $t \in [0, \check{t})$. Moreover, x is twice continuously differentiable due to $\Phi^{-1} \in C^2$.

Since $\bar{x}$ is a local minimal point of P, after possibly reducing $\check{t}$, all $t \in (0, \check{t})$ particularly fulfill

$$0 \leq f(x(t)) - f(\bar{x})$$

$$= t \langle \nabla f(\bar{x}), \dot{x}(0) \rangle + (t^2)/2 \left(\dot{x}(0)^{\mathsf{T}} D^2 f(\bar{x}) \dot{x}(0) + \langle \nabla f(\bar{x}), \ddot{x}(0) \rangle \right) + o(t^2).$$

Since for all $i \in I_0(\bar{x})$ we have $t\widetilde{y}_i \equiv 0$, it follows analogously

$$0 = \underbrace{g_i(x(t)) - g_i(\bar{x})}_{\equiv 0}$$

$$= t \langle \nabla g_i(\bar{x}), \dot{x}(0) \rangle + (t^2)/2 \left(\dot{x}(0)^{\mathsf{T}} D^2 g_i(\bar{x}) \dot{x}(0) + \langle \nabla g_i(\bar{x}), \ddot{x}(0) \rangle \right) + o(t^2)$$

as well as for all $j \in J$

$$0 = h_j(x(t)) - h_j(\bar{x})$$

$$= t \langle \nabla h_j(\bar{x}), \dot{x}(0) \rangle + (t^2)/2 \left(\dot{x}(0)^{\mathsf{T}} D^2 h_j(\bar{x}) \dot{x}(0) + \langle \nabla h_j(\bar{x}), \ddot{x}(0) \rangle \right) + o(t^2).$$

Multiplication of the equations with the corresponding multipliers and summing up yields

$$0 \leq (t^2)/2 \, \dot{x}(0)^{\mathsf{T}} D_x^2 L(\bar{x}, \bar{\lambda}, \bar{\mu}) \dot{x}(0) + o(t^2),$$

where we have used $\nabla_x L(\bar{x}, \bar{\lambda}, \bar{\mu}) = 0$. Division by t^2 and taking the limit $t \searrow 0$ results in

$$0 \leq \dot{x}(0)^{\mathsf{T}} D_x^2 L(\bar{x}, \bar{\lambda}, \bar{\mu}) \dot{x}(0) = d^{\mathsf{T}} D_x^2 L(\bar{x}, \bar{\lambda}, \bar{\mu}) d,$$

thus the assertion. $\qquad\square$

It can also be shown that $d^\mathsf{T} D_x^2 L(\bar{x}, \bar{\lambda}, \bar{\mu})d$ is nonnegative for d from the larger set $K(\bar{x}, M)$ [26].

Example 3.2.72 In Example 3.2.54, due to

$$D_x^2 L(\bar{x}, \bar{\mu})|_{T(\bar{x}, M)} = 1 - t$$

and Theorem 3.2.71, the critical point $\bar{x} = (0, 1)^\mathsf{T}$ of P is not a local minimal point of P for any $t > 1$. With the help of Theorem 3.2.67, it is easy to show that $\bar{x}$ is indeed a nondegenerate local *maximal* point of P for all $t > 1$. The sought-after critical parameter value is therefore $\bar{t} = 1$. Whether $\bar{x}$ is a local minimal or maximal point for $\bar{t} = 1$ cannot be clarified with the necessary and sufficient second-order conditions and must be examined separately if desired.

To conclude this example, we note that neither $D^2 f(\bar{x})$ nor $D_x^2 L(\bar{x}, \bar{\mu})$ are positive definite for any t.

3.2.10 Convex Optimization Problems

A constrained optimization problem

$$P: \quad \min\ f(x) \quad \text{s.t.} \quad x \in M$$

is called *convex*, if the set $M \subseteq \mathbb{R}^n$ and the function $f : M \to \mathbb{R}$ are convex.

Exercise 3.2.73 Let the functions $g_i : \mathbb{R}^n \to \mathbb{R}$, $i \in I$, be convex, and the functions $h_j : \mathbb{R}^n \to \mathbb{R}$, $j \in J$, be affine-linear, i.e., for all $j \in J$ it holds

$$h_j(x) = a_j^\mathsf{T} x + b_j$$

with some $a_j \in \mathbb{R}^n$ and $b_j \in \mathbb{R}$. Show that M is then a convex set.

Definition 3.2.74 (Convexly Described Set) We call a set given by inequality and equality constraints with arbitrary index sets I and J,

$$M = \left\{ x \in \mathbb{R}^n \,|\, g_i(x) \leq 0,\ i \in I,\ h_j(x) = 0,\ j \in J \right\},$$

convexly described, if the functions $g_i : \mathbb{R}^n \to \mathbb{R}$, $i \in I$, are convex and the functions $h_j : \mathbb{R}^n \to \mathbb{R}$, $j \in J$, are affine-linear.

In this terminology, Exercise 3.2.73 states that convexly described sets are convex.

In the following, let

$$A := \begin{pmatrix} a_1^\mathsf{T} \\ \vdots \\ a_q^\mathsf{T} \end{pmatrix} \quad \text{and} \quad b := \begin{pmatrix} b_1 \\ \vdots \\ b_q \end{pmatrix},$$

so $h(x) = Ax + b$. For the (q, n)-matrix A our basic assumption about the number of equality constraints implies $q < n$. Moreover, the gradients $\nabla h_1(\bar{x}) = a_1, \ldots, \nabla h_q(\bar{x}) = a_q$ are linearly independent if and only if $\text{rank}(A) = q$ holds.

For convex optimization problems with a convexly described feasible set, one may use a constraint qualification that involves only function values, but no gradients of nonlinear functions.

Definition 3.2.75 (Slater Constraint Qualification) Let the set $M \subseteq \mathbb{R}^n$ be convexly described. Then M satisfies the *Slater constraint qualification (SCQ)*, if the following two conditions are met:

(a) It holds $\text{rank}(A) = q$.
(b) There exists some point $x^\star \in \mathbb{R}^n$ with $g(x^\star) < 0$ and $h(x^\star) = 0$.

The SCQ is a *global* condition on M, because the *Slater point* $x^\star$ need not have anything to do with a minimal point of P. Indeed, *non*convex problems can possess a Slater point, while at the same time the MFCQ is violated at a minimal point. This is illustrated by the example $f(x) = x_1$, $g_1(x) = x_2 - x_1^3$, $g_2(x) = -x_2$ (Fig. 3.15).

On the other hand, in smooth convexly described optimization problems this effect cannot occur.

Fig. 3.15 SCQ without MFCQ at the minimal point

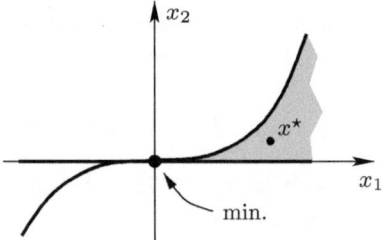

Theorem 3.2.76 *Let the set $M \subseteq \mathbb{R}^n$ be convexly described, nonempty, and let the functions g_i, $i \in I$, be continuously differentiable. Then the following statements are equivalent:*

(a) *The MFCQ holds everywhere in M.*
(b) *The MFCQ holds somewhere in M.*
(c) *M satisfies the SCQ.*

Proof We prove the assertion by showing the chain of implications 'statement a $\Rightarrow$ statement b $\Rightarrow$ statement c $\Rightarrow$ statement a'. Statement a implies statement b, because M is assumed to be nonempty.

To see that statement c follows from statement b, we choose a point $\bar{x} \in M$, at which the MFCQ holds. In particular, A then has rank q.

Let us first assume that $I_0(\bar{x})$ is empty. Then M satisfies the SCQ already with the Slater point $x^\star := \bar{x}$. For $I_0(\bar{x}) \neq \emptyset$ we choose some $d^0 \in \ker Dh(\bar{x}) = \ker A$ with $\langle \nabla g_i(\bar{x}), d^0 \rangle < 0$ for all $i \in I_0(\bar{x})$. The existence of such a d^0 is guaranteed by the MFCQ at $\bar{x}$. For each $i \in I_0(\bar{x})$ and all $t > 0$ it holds

$$g_i(\bar{x} + td^0) \;=\; \underbrace{g_i(\bar{x})}_{=\,0} + t\,\underbrace{\langle \nabla g_i(\bar{x}), d^0 \rangle}_{<\,0} + o(t),$$

so that scalars $t_i > 0$, $i \in I_0(\bar{x})$, with

$$g_i(\bar{x} + td^0) \;<\; 0 \quad \text{for all } t \in (0, t_i)$$

exist. With $t_0 = \min_{i \in I_0(\bar{x})} t_i > 0$ we thus obtain for all $t \in (0, t_0)$ and all $i \in I_0(\bar{x})$

$$g_i(\bar{x} + td^0) \;<\; 0.$$

The continuity of the functions g_i, $i \in I \setminus I_0(\bar{x})$, further guarantees

$$g_i(\bar{x} + td^0) \;<\; 0$$

for the inactive constraints, if $t \in (0, \check{t})$ holds with some sufficiently small $\check{t} \leq t_0$. Consequently, $g(x^\star) < 0$ holds for example with $x^\star := \bar{x} + (\check{t}/2)d^0$.

It remains to show $h(x^\star) = 0$. For all $j \in J$ it is

$$h_j(x^\star) = a_j^\mathsf{T}(\bar{x} + \tfrac{\check{t}}{2}d^0) + b_j \;=\; \underbrace{a_j^\mathsf{T}\bar{x} + b_j}_{=\,0} + \tfrac{\check{t}}{2}\,\underbrace{a_j^\mathsf{T}d^0}_{=\,0} \;=\; 0,$$

where the first summand vanishes because of $\bar{x} \in M$ and the second in view of $d^0 \in \ker A$. Thus, $x^\star$ is a Slater point of M.

Finally, let statement c hold, let $x^\star$ be a Slater point of M, and let $\bar{x} \in M$ be arbitrarily chosen. Because of $\mathrm{rank}(A) = q$ the gradients $\nabla h_1(\bar{x}), \dots, \nabla h_q(\bar{x})$ are linearly independent.

Furthermore, the vector $d^0 := x^\star - \bar{x}$ satisfies

$$Dh(\bar{x})d^0 \;=\; Ad^0 \;=\; (Ax^\star + b) - (A\bar{x} + b) \;=\; 0 - 0 \;=\; 0,$$

so that d^0 lies in the kernel of $Dh(\bar{x})$. Due to the C^1-characterization of convexity from Theorem 2.1.40, for each $i \in I_0(\bar{x})$ we also have

$$0 > g_i(x^\star) = g_i(\bar{x} + d^0) \geq \underbrace{g_i(\bar{x})}_{=\,0} + \langle \nabla g_i(\bar{x}), d^0 \rangle,$$

which shows $\langle \nabla g_i(\bar{x}), d^0 \rangle < 0$. Thus, at $\bar{x}$ the MFCQ is fulfilled.   □

That the *local* regularity condition MFCQ can be equivalent to the *global* regularity condition SCQ in the first place, is due to the assumption of a *global* structure on M, namely convexity.

According to Corollary 2.1.41, in the smooth convex *unconstrained* case, the critical points are identical with the global minimal points. One may therefore conjecture that, in the constrained case, the KKT points correspond to the global minimal points. In the following, we will see that this is at least correct under the SCQ. We call an optimization problem P *convexly described*, if M is convexly described and f is convex on M.

Corollary 3.2.77 *Let the problem P be convexly described, and let M satisfy the SCQ. Then every global minimal point of P is a KKT point.*

Proof Theorems 3.2.42 and 3.2.76.   □

Theorem 3.2.78 *Let $\bar{x}$ be a KKT point of the convexly described problem P. Then $\bar{x}$ is a global minimal point of P.*

Proof Let $\lambda_i \geq 0$, $i \in I_0(\bar{x})$, $\lambda_i = 0$, $i \in I \setminus I_0(\bar{x})$, and $\mu_j \in \mathbb{R}$, $j \in J$, be KKT multipliers for $\bar{x}$. Then all $x \in M$ satisfy

$$f(x) - f(\bar{x}) \overset{\text{Theorem 2.1.40}}{\geq} \langle \nabla f(\bar{x}), x - \bar{x} \rangle$$

$$\overset{\bar{x}\ \text{KKT point}}{=} -\sum_{i \in I} \lambda_i \langle \nabla g_i(\bar{x}), x - \bar{x} \rangle - \sum_{j \in J} \mu_j \langle \nabla h_j(\bar{x}), x - \bar{x} \rangle$$

$$\overset{\text{Theorem 2.1.40},\ \lambda \geq 0}{\geq} -\sum_{i \in I} \lambda_i (g_i(x) - g_i(\bar{x})) - \sum_{j \in J} \mu_j (\underbrace{h_j(x)}_{=\,0} - \underbrace{h_j(\bar{x})}_{=\,0})$$

$$\overset{\lambda_{I_0^c} = 0}{=} -\sum_{i \in I_0(\bar{x})} \underbrace{\lambda_i}_{\geq 0}\ \underbrace{(g_i(x)}_{\leq 0} - \underbrace{g_i(\bar{x}))}_{=\,0}$$

$$\geq 0.$$

                                                                                          □

In contrast to the necessary optimality condition from Corollary 3.2.77, the sufficient condition in Theorem 3.2.78 does not require any regularity condition. Overall, we obtain the following 'characterization' for global minimal points of convexly described problems P:

$$\bar{x} \text{ global minimal point} \overset{\text{SCQ}}{\Rightarrow} \bar{x} \text{ KKT point} \Rightarrow \bar{x} \text{ global minimal point.}$$

We remark that such a characterization of global minimal points by KKT points even holds in the 'hidden convex case' where a convex set M is described by inequality constraints with nonconvex C^1-functions $g_i, i \in I$, as long as the SCQ and a mild nondegeneracy assumption are satisfied [24]. The proof of this result partly relies on concepts from convex analysis and would thus go beyond the scope of this textbook.

For a presentation of the duality theory of convex optimization problems, we refer to [36].

3.3 Algorithms

This section discusses some algorithms for solving the problem

$$P: \quad \min_{x \in \mathbb{R}^n} f(x) \quad \text{s.t.} \quad g_i(x) \le 0, \, i \in I, \quad h_j(x) = 0, \, j \in J,$$

with sufficiently smooth functions f, g, and h. We essentially distinguish two classes of algorithms for P:

- The successive approximation of P by unconstrained optimization problems is the basic idea of the penalty method (Sect. 3.3.1), the multiplier method (Sect. 3.3.2), and the barrier method (Sect. 3.3.3). Such methods are called *primal*.
- *Primal-dual* methods, on the other hand, attempt to directly solve the Karush-Kuhn-Tucker system of P. These include primal-dual interior point methods (Sect. 3.3.4) and the SQP method (Sect. 3.3.5).

In the optimization literature also the multiplier method is often considered primal-dual, although it does not aim at solving the KKT system. The reason will become apparent below. In convex optimization, there are further algorithmic approaches that approximate the original convex problem by linear optimization problems and solve these using, e.g., the simplex algorithm (e.g., Kelley's cutting plane method and the Frank-Wolfe method). Given that nowadays smooth convex problems are 'easy' to solve, an analogous approach in global optimization is to approximate an original nonconvex problem by a series of convex problems. For a detailed presentation of such methods, we refer to [36].

We dedicate the separate Sect. 3.3.6 to the step size control in the SQP method, as this should not only guarantee a descent in the objective function value, but also 'improved feasibility' in a certain sense. This can be implemented, for example, by merit functions or filters. The concluding Sect. 3.3.7 presents a way to solve quadratic optimization problems, as they occur in projection problems to polyhedra or as auxiliary problems, for example, in the SQP method and in trust-region methods. For some algorithms we limit ourselves to the presentation of their main ideas, and refer to [2, 26] for their pseudocodes.

3.3.1 Penalty Methods

The basic idea of penalty methods is to formulate the feasibility of a point (i.e., the requirement $x \in M$) as a second objective, in addition to minimizing the function f. The following concept is used for this purpose.

Definition 3.3.1 (Penalty Function) A function $\alpha : \mathbb{R}^n \to \mathbb{R}$ is called *penalty function* for $M \subseteq \mathbb{R}^n$, if the following conditions are met:

(a) All $x \in M$ satisfy $\alpha(x) = 0$.
(b) All $x \in M^c$ satisfy $\alpha(x) > 0$.

The values of a penalty function can be interpreted as a 'penalty' for the infeasibility of a point x. Because of $\alpha(x) \geq 0$ for all $x \in \mathbb{R}^n$ and $\alpha(x) = 0$ for all $x \in M$, the elements of M coincide with the global minimal points of the auxiliary function α. Thus, the requirement that x should be a feasible point is reformulated as a second objective. One can therefore try to solve the unconstrained *multicriteria problem*

$$\min_x \begin{pmatrix} f(x) \\ \alpha(x) \end{pmatrix}$$

instead of P. What is understood by a solution of a multicriteria problem is explained, for example, in [25].

One popular approach for the solution of multicriteria problems is *scalarization* by minimizing a *weighted sum* of the objective functions. Indeed, the penalty method takes this approach and minimizes a weighted sum of the two objective functions f and α. Due to Exercise 1.3.1a it is sufficient to weight only one of the two functions. This leads to the approach of replacing P with an unconstrained problem with objective function

$$A(t, x) = f(x) + t \cdot \alpha(x)$$

and a suitable parameter $t > 0$, i.e., by

$$P(t): \quad \min_{x \in \mathbb{R}^n} f(x) + t \cdot \alpha(x).$$

The *exact* solution of P would be obtained for each t with the 'ideal' penalty function which satisfies $\alpha(x) = +\infty$ for all $x \in M^c$. Since this is neither feasible nor manageable algorithmically, other penalty functions are constructed.

With the positive part

$$g_i^+(x) := \max\{0, g_i(x)\}, \quad i \in I,$$

of $g_i(x)$ we have $x \in M$ if and only if the vector

$$(g_1^+(x), \ldots, g_p^+(x), h_1(x), \ldots, h_q(x))$$

vanishes entrywise. Due to the definiteness of norms, the latter is the case if and only if, with some $r \in \mathbb{N}$, the term

$$\alpha_r(x) := \|(g_1^+(x), \ldots, g_p^+(x), h_1(x), \ldots, h_q(x))\|_r^r$$

disappears (i.e., the r-th power of the ℓ_r-norm). Because of $\alpha_r(x) > 0$ for all $x \in M^c$, α_r is a penalty function for M. In the following, we will pay special attention to the penalty functions α_1 and α_2. The ℓ_1-*penalty function*

$$\alpha_1(x) = \sum_{i \in I} g_i^+(x) + \sum_{j \in J} |h_j(x)|$$

is continuous and one-sided directionally differentiable, but not differentiable.

Exercise 3.3.2 Show with the help of Exercise 2.1.12, that the functions $\varphi(a) = \max\{0, a\}$ and $\text{abs}(a) = |a|$ are one-sided directionally differentiable at every $a \in \mathbb{R}$, with

$$\varphi'(a, d) = \begin{cases} 0, & \text{if } a < 0 \\ \max\{0, d\}, & \text{if } a = 0 \\ d, & \text{if } a > 0 \end{cases}$$

and

$$\text{abs}'(a, d) = \begin{cases} -d, & \text{if } a < 0 \\ |d|, & \text{if } a = 0 \\ d, & \text{if } a > 0 \end{cases}$$

for all $d \in \mathbb{R}$.

Exercise 3.3.3 Let

$$I_-(x) = \{i \in I \mid g_i(x) < 0\},$$
$$I_0(x) = \{i \in I \mid g_i(x) = 0\},$$
$$I_+(x) = \{i \in I \mid g_i(x) > 0\},$$
$$J_-(x) = \{j \in J \mid h_j(x) < 0\},$$
$$J_0(x) = \{j \in J \mid h_j(x) = 0\},$$
$$J_+(x) = \{j \in J \mid h_j(x) > 0\}.$$

Show for continuously differentiable functions g and h on $\mathbb{R}^n$ the one-sided directional differentiability of the ℓ_1-penalty function at every $x \in \mathbb{R}^n$ with

$$\alpha_1'(x, d) = \sum_{i \in I_0(x)} \max\{0, \langle \nabla g_i(x), d \rangle\} + \sum_{i \in I_+(x)} \langle \nabla g_i(x), d \rangle$$
$$- \sum_{j \in J_-(x)} \langle \nabla h_j(x), d \rangle + \sum_{j \in J_0(x)} |\langle \nabla h_j(x), d \rangle| + \sum_{j \in J_+(x)} \langle \nabla h_j(x), d \rangle$$

for all $d \in \mathbb{R}^n$.

In contrast, the ℓ_2-*penalty function*

$$\alpha_2(x) = \sum_{i \in I} (g_i^+(x))^2 + \sum_{j \in J} (h_j(x))^2$$

is continuously differentiable on $\mathbb{R}^n$.

Exercise 3.3.4 Show that the function $\varphi^2(a) = (\max\{0, a\})^2$ is continuously differentiable at every $a \in \mathbb{R}$ and that for continuously differentiable functions g and h on $\mathbb{R}^n$, α_2 is also continuously differentiable on all of $\mathbb{R}^n$.

In the following, for $r \in \{1, 2\}$ we denote with

$$P_r(t): \quad \min_{x \in \mathbb{R}^n} f(x) + t \cdot \alpha_r(x)$$

the unconstrained optimization problem formed with the penalty function α_r.

Example 3.3.5 For $g(x) = x$ we have $g^+(x) = \max\{0, x\}$, which implies $\alpha_1(x) = \max\{0, x\}$ and $\alpha_2(x) = (\max\{0, x\})^2$. The two penalty functions are sketched in Fig. 3.16.

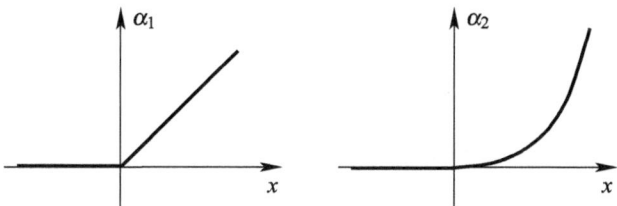

Fig. 3.16 ℓ_1- and ℓ_2-penalty functions

Fig. 3.17 Problems $P_2(t)$
with various $t > 0$

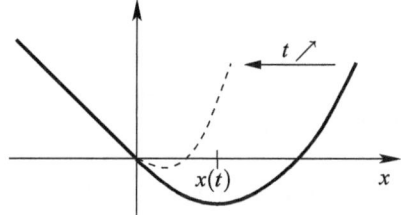

With the objective function $f(x) = -x$, the minimal point of P is $\bar{x} = 0$. The ℓ_2-penalty function leads to the family of unconstrained problems

$$P_2(t): \quad \min_{x \in \mathbb{R}^n} -x + t(\max\{0, x\})^2$$

with for P *infeasible* minimal points $x(t)$, $t > 0$ (Fig. 3.17). However, one can see from the figure that $\lim_{t \to \infty} x(t) = 0 = \bar{x}$ holds, i.e., the minimal points of $P_2(t)$ at least converge for increasing penalty parameters t to the exact minimal point of P.

In fact, the following result applies in general.

Theorem 3.3.6 *Let the function* $\alpha : \mathbb{R}^n \to \mathbb{R}$ *be a continuous penalty function for M, let* (t^k) *be a monotonically increasing sequence with* $\lim_k t^k = +\infty$, *and for all* $k \in \mathbb{N}$ *let* x^k *be a global minimal point of* $P(t^k)$. *Then every accumulation point* $x^\star$ *of* (x^k) *is a global minimal point of P.*

Proof Let $x^\star$ be an accumulation point, and let us assume $\lim_k x^k = x^\star$ to avoid a subsequence notation. Also, let $x \in M$ be arbitrary. Due to the nonnegativity of t^k and $\alpha(x^k)$, and due to the minimality property of x^k, for all $k \in \mathbb{N}$ we obtain

$$f(x^k) \leq f(x^k) + t^k \alpha(x^k) \leq f(x) + t^k \underbrace{\alpha(x)}_{= 0}.$$

From this, firstly

$$f(x^\star) = \lim_k f(x^k) \leq f(x) \tag{3.21}$$

follows and, secondly, for all $k \in \mathbb{N}$

$$0 \leq t^k \alpha(x^k) \leq \underbrace{f(x) - f(x^k)}_{\to\ f(x) - f(x^\star)}. \tag{3.22}$$

Therefore, the sequence $(t^k \alpha(x^k))$ is bounded. In view of $t^k \to \infty$ this is only possible for $\lim_k \alpha(x^k) = 0$. Because of the continuity of α, it also holds $\alpha(x^\star) = \lim_k \alpha(x^k)$, so that $x^\star$ must lie in M by definition of a penalty function. Together with (3.21) this proves the assertion.   □

A possible implementation of the penalty method is given in Algorithm 3.2. The termination criterion takes advantage of the fact that $\alpha(x)$ can be considered a 'measure' of the infeasibility of x and that for $k \to \infty$ not only $\alpha(x^k)$, but even $t^k \alpha(x^k)$ tends to zero. The latter follows from (3.22) with the special choice $x = x^\star$.

Algorithm 3.2: Penalty method

Input: Solvable C^1-optimization problem P, penalty function α, starting point $x^0 \in \mathbb{R}^n$,
            starting parameter $t^0 > 0$, factor $\rho > 1$ and termination tolerance $\varepsilon > 0$
Output: Approximation $\bar{x}$ of a global minimal point of P (if the method terminates;
            Theorem 3.3.6)

1 **begin**
2   Set $k = 0$.
3   **repeat**
4   Replace k with $k + 1$.
5   Set $t^k = \rho\, t^{k-1}$.
6   Starting from x^{k-1}, determine a global minimal point x^k of

$$P(t^k): \quad \min\ f(x) + t^k\, \alpha(x).$$

7   **until** $t^k \alpha(x^k) < \varepsilon$
8   Set $\bar{x} = x^k$.
9 **end**

Algorithm 3.2 is popular in practice due to its simple implementability. However, it is only conceptual in the sense that the methods from Sect. 2.2 cannot be expected to identify *global* minimal points of the auxiliary problems $P(t^k)$ from line 6 without further assumptions. The applicability of the methods from Sect. 2.2 also requires a *smooth* penalty function, for example α_2, but not α_1.

Fig. 3.18 Problems $P_1(t)$
with various $t > 0$

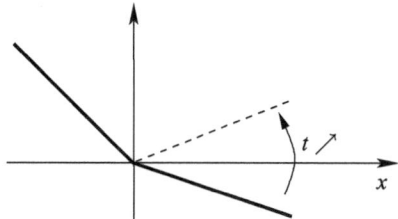

Even when using a smooth penalty function, the numerical treatment of the problems $P(t)$ becomes increasingly difficult for growing t, as the objective function $f(x) + t\alpha(x)$ then exhibits an increasingly strong curvature at the boundary of M. If a certain threshold for t is exceeded, the theoretically smooth objective function of $P(t)$ is numerically no longer distinguishable from a function that is nonsmooth at the boundary of M, so that the methods from Sect. 2.2 may no longer work.

Another disadvantage of penalty methods is that the points $x(t)$ with $t > 0$ must be expected to be *infeasible* for P. Example 3.3.5 also illustrates that with the differentiable ℓ_2-penalty term an exact minimal point of P is generally only obtained for $t \to \infty$. From Algorithm 3.2 one can therefore only expect to generate a 'nearly feasible' point x^k in the sense that its feasibility measure satisfies $\alpha(x^k) < \varepsilon/t^k$, instead of $\alpha(x^k) = 0$.

With the ℓ_1-penalty function α_1, on the other hand, one finds the exact minimal point of P under mild conditions already for *sufficiently large t*, which we will first convince ourselves of using an example.

Example 3.3.7 The penalty term α_1 in Example 3.3.5 leads to the nonsmooth unconstrained problem

$$P_1(t): \quad \min_{x \in \mathbb{R}^n} -x + t \max\{0, x\}$$

(Fig. 3.18).

For all $t > 1$, $x(t) = 0$ is the exact minimal point of P. The method with ℓ_1-penalty function is therefore called *exact penalty method*.

Theorem 3.3.8 *Let $\bar{x}$ be a nondegenerate local minimal point of P with multipliers $\bar{\lambda}$, $\bar{\mu}$. Then for all*

$$t > \|(\bar{\lambda}_{I_0}, \bar{\mu})\|_\infty \ (= \max\{\bar{\lambda}_i, \ i \in I_0(\bar{x}), \ |\bar{\mu}_j|, \ j \in J\})$$

$\bar{x}$ is also a local minimal point of $P_1(t)$.

Proof Since at a nondegenerate local minimal point active inequality constraints act like equality constraints, it is sufficient to examine the case $I_0(\bar{x}) = \emptyset$. With the parametric auxiliary problem

$$P(s): \quad \min_{x \in \mathbb{R}^n} f(x) \quad \text{s.t.} \quad \widetilde{h}_j(s, x) := h_j(x) - s_j = 0, \quad j \in J,$$

and parameter vector $s \in \mathbb{R}^q$, $\bar{x}$ is a nondegenerate local minimal point of $P(0)$. In [34], it is shown that there then exists a locally unique function $x(s)$ defined locally around $\bar{s} = 0$ with $x(\bar{s}) = \bar{x}$, such that $x(s)$ is a nondegenerate local minimal point of $P(s)$. The associated critical value function $\bar{v}(s) = f(x(s))$ is therefore twice continuously differentiable at $\bar{s} = 0$ and also satisfies $\nabla \bar{v}(0) = -\bar{\mu}$. For neighborhoods U of $\bar{x}$ and V of $\bar{s} = 0$, it follows

$$\min_{x \in U} f(x) + t \cdot \alpha_1(x) = \min_{x \in U} f(x) + t \sum_{j \in J} |h_j(x)| \tag{3.23}$$

$$= \min_{(s,x) \in V \times U} \left\{ f(x) + t \sum_{j \in J} |s_j| \,\Big|\, h_j(x) = s_j, \ j \in J \right\}$$

$$= \min_{s \in V} \left[\min_{x \in U} \left\{ f(x) \,\Big|\, \widetilde{h}_j(s, x) = 0, \ j \in J \right\} + t \sum_{j \in J} |s_j| \right]$$

$$= \min_{s \in V} \bar{v}(s) + t \sum_{j \in J} |s_j|. \tag{3.24}$$

Furthermore, all $s \in V$ satisfy

$$\bar{v}(s) + t \sum_{j \in J} |s_j| = \bar{v}(0) + \langle \nabla v(0), s \rangle + o(\|s\|) + t \sum_{j \in J} |s_j| \tag{3.25}$$

$$= f(\bar{x}) + \underbrace{\langle -\bar{\mu}, s \rangle}_{\geq -\sum_{j \in J} |\bar{\mu}_j| \cdot |s_j|} + o(\|s\|) + t \sum_{j \in J} |s_j|$$

$$\geq f(\bar{x}) + \sum_{j \in J} (t - |\bar{\mu}_j|) |s_j| + o(\|s\|). \tag{3.26}$$

By assumption, $t - |\bar{\mu}_j| > 0$ holds for all $j \in J$, so that

$$n(s) := \sum_{j \in J} (t - |\mu_j|) |s_j|$$

is a norm on $\mathbb{R}^q$. We therefore choose for the norm in the Taylor remainder term $\|s\| := n(s)$ and can thus bound the expression

$$f(\bar{x}) + \|s\|(1 + o(1))$$

from (3.26) for sufficiently small $\|s\|$ by $f(\bar{x})$ from below. After possibly shrinking V, and in view of $\bar{v}(0) + t \sum_{j \in J} |0| = f(\bar{x})$, this implies

$$\min_{s \in V} \bar{v}(s) + t \sum_{j \in J} |s_j| = f(\bar{x}).$$

Together with (3.24), this is the assertion. $\square$

A significant disadvantage of the ℓ_1-penalty approach is the lacking smoothness of the unconstrained auxiliary problems $P_1(t)$, $t > 0$, so that this method is predominantly used for problems where P itself is already nonsmooth (see however [26] for exceptions).

Another disadvantage is that the limit $\|(\bar\lambda_{I_0}, \bar\mu)\|_\infty$ is not known a priori. In Sect. 3.3.6 we will still be able to use the ℓ_1-penalty function in an algorithmically meaningful way.

3.3.2 Multiplier Method

A smooth *and* exact penalty term is provided by the following consideration, which we again only present for the case $I = \emptyset$, as active inequality constraints at nondegenerate critical points behave like equality constraints.

To this end, we first briefly discuss the basic idea of the *projected gradient method* for the case $I = \emptyset$. Consider a point $\bar x \in M$ at which the LICQ is fulfilled. Then it is to be expected (Exercise 3.3.9), that at $\bar x$ a feasible first-order descent direction can be generated by *orthogonal projection* of the negative gradient $-\nabla f(\bar x)$ onto the outer linearization cone $L_\leq(\bar x, M)$, i.e., the direction $\bar d = \mathrm{pr}(-\nabla f(\bar x), L_\leq(\bar x, M))$ is calculated. It is defined as the unique minimal point of the problem

$$Q: \quad \min_d \tfrac{1}{2}\|d + \nabla f(\bar x)\|_2^2 \quad \text{s.t.} \quad \langle \nabla h_j(\bar x), d \rangle = 0, \ j \in J.$$

Since Q is convex and has linear constraints, according to Corollary 3.2.24 and Theorem 3.2.78 its global minimal points coincide with its KKT points, i.e., with the $\bar d$-part of any solution $(\bar d, \bar\mu)$ of

$$d + \nabla f(\bar x) + \sum_{j \in J} \mu_j \nabla h_j(\bar x) = 0,$$

$$\langle \nabla h_j(\bar x), d \rangle = 0, \quad j \in J.$$

In particular, without further calculation of $\bar\mu$ one obtains $\bar d = -\nabla_x L(\bar x, \bar\mu)$.

Exercise 3.3.9 Show that the solvability of Q by $\bar d = 0$ is equivalent to $\bar x$ being a KKT point of P and that every minimal point $\bar d \neq 0$ of Q is a first-order descent direction for f in $\bar x$.

These observations do not yet lead to an algorithm, since points of the form $\bar x + t\bar d$ with t from some step size control are not necessarily feasible if the functions h_j, $j \in J$, are nonlinear. Projected gradient methods, for whose details we refer to [7,26], make the necessary adjustments.

We rather employ the fact that the search direction $\bar{d} = -\nabla_x L(\bar{x}, \bar{\mu})$ can as well be interpreted as the gradient descent direction for the unconstrained objective function $L(\cdot, \mu)$ with suitably chosen multiplier $\mu := \bar{\mu}$. This motivates to pursue the approach of minimizing the function $L(\cdot, \mu)$ over M via the ℓ_2-penalty method, which explains the two synonyms *multiplier method* and *augmented Lagrangian method* for the procedure to be discussed. The resulting auxiliary problem depends on the penalty parameter t and on the guess μ for the multiplier $\bar{\mu}$ in the solution point $\bar{x}$:

$$P(t, \mu): \quad \min_{x \in \mathbb{R}^n} f(x) + \sum_{j \in J} \mu_j h_j(x) + t \sum_{j \in J} h_j^2(x).$$

The basic idea of the method is therefore that the minimal point $\bar{x}$ of P possesses some KKT multiplier $\bar{\mu}$, and that an estimate μ^k for it is available. For $t^k > 0$, let x^k denote a local minimal point of $P(t^k, \mu^k)$. Then x^k satisfies the first-order necessary optimality condition

$$0 = \nabla_x \left(f(x) + \sum_{j \in J} \mu_j^k h_j(x) + t^k \sum_{j \in J} h_j^2(x) \right) |_{x = x^k}$$

$$= \nabla f(x^k) + \sum_{j \in J} \mu_j^k \nabla h_j(x^k) + t^k \sum_{j \in J} 2 h_j(x^k) \nabla h_j(x^k)$$

$$= \nabla f(x^k) + \sum_{j \in J} \underbrace{(\mu_j^k + 2 t^k h_j(x^k))}_{=: \, \mu_j^{k+1}} \nabla h_j(x^k).$$

The value μ_j^{k+1} is used as an update of μ^k for the next auxiliary problem $P(t^{k+1}, \mu^{k+1})$. For μ^k sufficiently close to $\bar{\mu}$, the statement of the next result can be transferred to $P(t^k, \mu^k)$.

Theorem 3.3.10 *Let $\bar{x}$ be a nondegenerate local minimal point of P with multiplier $\bar{\mu}$. Then there exists some $\bar{t} > 0$, such that $\bar{x}$ is a strict local minimal point of $\pi(x, t, \bar{\mu}) := L(x, \bar{\mu}) + t \|h(x)\|_2^2$ for all $t > \bar{t}$.*

Proof Because of

$$\nabla_x \pi(x, t, \bar{\mu}) = \nabla_x L(x, \bar{\mu}) + 2t \sum_{j \in J} h_j(x) \nabla h_j(x)$$

$\bar{x}$ is a critical point of $\pi(\cdot, t, \bar{\mu})$ for all $t > 0$. We show that in addition some $\bar{t} > 0$ exists, such that $D_x^2 \pi(\bar{x}, t, \bar{\mu})$ is positive definite for all $t > \bar{t}$.

Assume that this is not the case. Then for all $k \in \mathbb{N}$ there exist some $t^k > k$ and $d^k \in B_=(0, 1)$ with

$$(d^k)^\mathsf{T} D_x^2 \pi(\bar{x}, t^k, \bar{\mu}) \, d^k \; \leq \; 0. \tag{3.27}$$

Due to

$$D_x^2 \pi(\bar{x}, t, \bar{\mu}) \; = \; D_x^2 L(\bar{x}, \bar{\mu}) + 2t \sum_{j \in J} \big(\nabla h_j(\bar{x}) D h_j(\bar{x}) + \underbrace{h_j(\bar{x})}_{=0} D_x^2 h_j(\bar{x}) \big),$$

Equation (3.27) is equivalent to

$$(d^k)^\mathsf{T} D_x^2 L(\bar{x}, \bar{\mu}) \, d^k + 2t^k \sum_{j \in J} \big(\langle \nabla h_j(\bar{x}), d^k \rangle \big)^2 \; \leq \; 0. \tag{3.28}$$

After possibly transitioning to a subsequence, $d^k \to \bar{d} \in B_=(0, 1)$ holds, and thus $(d^k)^\mathsf{T} D_x^2 L(\bar{x}, \bar{\mu}) d^k \to \bar{d}^\mathsf{T} D_x^2 L(\bar{x}, \bar{\mu}) \bar{d} \in \mathbb{R}$. Because of $t^k \to \infty$, (3.28) implies

$$\sum_{j \in J} \langle \nabla h_j(\bar{x}), d^k \rangle^2 \; \to \; 0,$$

thus for all $j \in J$

$$0 \; = \; \lim_k \langle \nabla h_j(\bar{x}), d^k \rangle \; = \; \langle \nabla h_j(\bar{x}), \bar{d} \rangle$$

and $\bar{d} \in T(\bar{x}, M)$. Overall, according to (3.28) we have constructed some $\bar{d} \in T(\bar{x}, M)$ with $\bar{d}^\mathsf{T} D_x^2 L(\bar{x}, \bar{\mu}) \bar{d} \leq 0$, which contradicts $D_x^2 L(\bar{x}, \bar{\mu})|_{T(\bar{x}, M)} \succ 0$. □

That the limit to be exceeded by t is unknown a priori, is a disadvantage which the multiplier method shares with the exact penalty method. For practical implementations of this method, we refer to [26].

We note that the function $\pi(x, t, \mu) := L(x, \mu) + t \|h(x)\|_2^2$, used as the objective function of the multiplier method, also plays a central role in modern developments of duality theory [28].

3.3.3 Barrier Methods

Barrier methods treat the inequality constraints in P, so in their following motivation we focus on the case without equality constraints. In the case $J \neq \emptyset$, the barrier approach at least reduces P to a problem with $I = \emptyset$.

For the problem

$$P: \qquad \min_{x \in \mathbb{R}^n} \; f(x) \quad \text{s.t.} \quad g_i(x) \leq 0, \; i \in I,$$

let

$$M_< := \{x \in \mathbb{R}^n \mid g_i(x) < 0, \ i \in I\} \neq \emptyset,$$

i.e., the Slater constraint qualification (SCQ) known from convex optimization is fulfilled. The basic idea of barrier methods is to replace P with a problem in which the inequality constraints cannot become active. In the following, we denote by bd$A$ the (topological) *boundary* of a set $A \subseteq \mathbb{R}^n$ (i.e., the set of all points for which each of its neighborhoods contains both an element of A and an element of A^c).

Definition 3.3.11 (Barrier Function) The function $\beta : M_< \to \mathbb{R}$ is called a *barrier function* for M, if for all sequences $(x^k) \subseteq M_<$ with $\lim_k x^k = \bar{x} \in$ bd$(M_<)$

$$\lim_k \beta(x^k) = +\infty$$

holds.

The defining property of the barrier function can be interpreted as a 'penalty for being too close to the boundary of M'. An important barrier term is (Frisch's) *logarithmic barrier function*

$$\beta(x) = -\sum_{i \in I} \log\left(-g_i(x)\right).$$

A barrier method replaces P with the problem

$$P(t): \quad \min_{x \in \mathbb{R}^n} \ f(x) + t \cdot \beta(x) \quad \text{s.t.} \quad x \in M$$

with $t > 0$, where the inequality constraints describing the set M cannot become active due to the barrier term.

The objective function of $P(t)$,

$$B(t, x) := f(x) + t\beta(x),$$

is only defined on $M_<$. For $t \searrow 0$, all points in $M_<$, including those near bd$(M_<)$, are penalized successively less, as for fixed $x \in M_<$ the expressions $t\beta(x)$ converge pointwise to zero. The proof of Theorem 3.3.13 will show how this argument can be extended to boundary points of $M_<$, to yield a convergence result for minimal points.

Fig. 3.19 Logarithmic
barrier function β

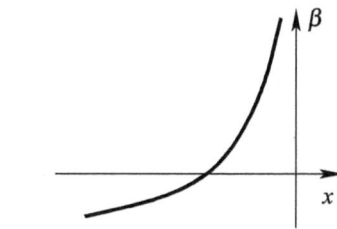

Fig. 3.20 Problems $P(t)$
with various $t > 0$

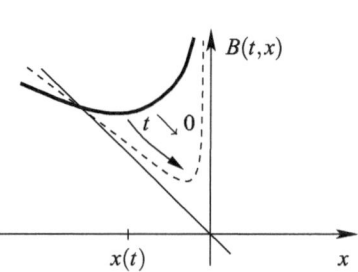

Example 3.3.12 As in Examples 3.3.5 and 3.3.7 we consider $f(x) = -x$ and $g(x) = x$.
We obtain the barrier term sketched in Fig. 3.19 and the unconstrained problems
shown in Fig. 3.20. As opposed to penalty methods, their minimal points $x(t)$, $t > 0$,
are feasible for P. For $t \searrow 0$, the minimal points of $B(t, x)$ converge to $\bar{x} = 0$, the
exact minimal point of P.

Indeed, the following result is true.

Theorem 3.3.13 *Assume $M = \mathrm{cl}(M_<)$, let β be continuous on $M_<$, and for
$t^k \searrow 0$ let x^k be global minimal points of $P(t^k)$. Then every accumulation
point $x^\star$ of (x^k) is a global minimal point of P.*

Proof Let $x^\star$ be an accumulation point, and let us assume without loss of generality
$\lim_k x^k = x^\star$. Because of $x^k \in M_<$ for all $k \in \mathbb{N}$, $x^\star$ lies in $\mathrm{cl}(M_<) = M$. Therefore,
we only need to show $f(x^\star) - f(x) \leq 0$ for every $x \in M$.

Case 1: $x \in M_<$. Since x^k is a global minimal point of $P(t^k)$ for all $k \in \mathbb{N}$, it holds

$$f(x^k) + t^k \beta(x^k) \leq f(x) + t^k \beta(x)$$

and hence

$$f(x^k) - f(x) \leq t^k \left(\beta(x) - \beta(x^k) \right). \tag{3.29}$$

Since the left side of (3.29) converges to $f(x^\star) - f(x)$ for $k \to \infty$, the desired result is shown if we can prove that the right side either converges to zero or can be bounded above by zero.

Case 1.1: $x^\star \in M_<$. The continuity of β yields $\beta(x^k) \to \beta(x^\star)$, from which with (3.29) and $\lim_k t^k = 0$ the assertion $f(x^\star) - f(x) \le 0$ follows.

Case 1.2: $x^\star \in \mathrm{bd}(M_<)$. According to the definition of a barrier function, it holds $\lim_k \beta(x^k) = +\infty$, so that $\beta(x) - \beta(x^k) \le 0$ is guaranteed for all sufficiently large k. For these k, from the positivity of the t^k and (3.29), $f(x^k) - f(x) \le 0$ follows, thus in the limit $f(x^\star) - f(x) \le 0$.

Case 2: $x \in M \setminus M_<$. In this case, $x \in \mathrm{cl}(M_<) \setminus M_< = \mathrm{bd}(M_<)$ holds, so that there exists a sequence $x^\ell \to x$ with $(x^\ell) \subseteq M_<$. According to the first case, $f(x^\star) \le f(x^\ell)$ is true for all $\ell \in \mathbb{N}$, thus in the limit $f(x^\star) \le f(x)$. $\square$

Exercise 3.3.14 Show that the condition $M = \mathrm{cl}(M_<)$ from Theorem 3.3.13 can be guaranteed by requiring the MFCQ to hold at every point of M.

A conceptual implementation of the barrier method is given in Algorithm 3.3. For the algorithmic treatment of the auxiliary problems $P(t^k)$ in Algorithm 3.3 by the methods from Sect. 2.2, it is necessary to choose a barrier function β that is at least continuously differentiable on $M_<$. However, like Algorithm 3.2, also Algorithm 3.3 is only conceptual, because without further assumptions the methods from Sect. 2.2 cannot be expected to identify *global* minimal points of the auxiliary problems $P(t^k)$ in line 6.

As a further analogy, at least to the ℓ_2-penalty method, the numerical treatment of the problems $P(t^k)$ becomes increasingly difficult for decreasing t^k, as the barrier function β then exhibits an increasingly strong curvature close to the boundary of M (Exercise 3.3.16). Also, when algorithmically solving the auxiliary problem $P(t^k)$ with one of the methods from Sect. 2.2, one must keep in mind that a too large step size can lead to the violation of the theoretically inactive constraints $g_i(x) \le 0$, $i \in I$. Therefore, the constraints must be explicitly considered algorithmically and, if necessary, sufficiently small step sizes need to be computed.

At least, for all $t > 0$ every minimal point of $P(t)$ is an interior point of M, so barrier methods are (primal) *interior point methods*. If one has to terminate an interior-point method after a few steps without having met a termination criterion, at least a feasible point with usually improved objective function value has been computed (see also Remark 2.2.2).

The fact that Algorithm 3.3 is not commonly used in the form presented, is primarily due to the simple adjustment of the barrier parameter in line 5 in combination with the simple termination criterion in line 7. Indeed, from the knowledge of t^0, ρ and ε one could a priori calculate the number of required iterations. Under additional conditions the alternative termination criterion $|t^k \beta(x^k)| < \varepsilon$ could be used [2], but let us rather focus on the crucial modification of Algorithm 3.3 to

---

Algorithm 3.3: Barrier method

Input: Solvable C^1-optimization problem P with $J = \emptyset$ and $M_< \neq \emptyset$,
             barrier function β, starting point $x^0 \in M_<$, starting parameter $t^0 > 0$,
             factor $\rho \in (0, 1)$ and termination tolerance $\varepsilon > 0$

Output: Approximation $\bar{x}$ of a global minimal point of P (if the method terminates;
             Theorem 3.3.13)

---

1 **begin**
2   Set $k = 0$.
3   **repeat**
4   Replace k with $k + 1$.
5   Set $t^k = \rho\, t^{k-1}$.
6   Starting from x^{k-1}, determine a global minimal point x^k of

$$P(t^k): \quad \min\ f(x) + t^k\,\beta(x) \quad \text{s.t.} \quad x \in M.$$

7   **until** $t^k < \varepsilon$
8   Set $\bar{x} = x^k$.
9 **end**

---

a practically relevant procedure. This is its modification to a *primal-dual interior point method*, which we will discuss in the next section. There, on the one hand, the barrier parameter t^k is adjusted more cleverly, and on the other hand a deeper interpretation of the termination criterion from line 7 arises.

This interpretation of the termination criterion $t^k < \varepsilon$ is possible when Algorithm 3.3 or its variation from the following section is applied to a C^1-optimization problem P with *convex* functions f and g_i, $i \in I$, as well as with the *logarithmic barrier function* β.

To this end, we set up the Wolfe-dual problem of P, derived for example in [36],

$$D: \quad \max_{x,\lambda} L(x, \lambda) \quad \text{s.t.} \quad \nabla_x L(x, \lambda) = 0, \ \lambda \geq 0,$$

where L again denotes the Lagrange function of P. Furthermore, let v_D be the supremum of the function L over the feasible set of D, and v_P be the infimum of the function f over the feasible set of P. Then, for every convexly described C^1-problem P, the weak duality theorem [36] ensures $v_D \leq v_P$. The nonnegative difference $v_P - v_D$ is called *duality gap*.

By definition of v_P and v_D, for every feasible point $\bar{x}$ of P and every $\bar{\lambda}$ with $(\bar{x}, \bar{\lambda})$ feasible for D, we have

$$f(\bar{x}) - L(\bar{x}, \bar{\lambda}) \ \geq \ v_P - v_D \ \geq 0,$$

so that the difference $f(\bar{x}) - L(\bar{x}, \bar{\lambda})$ provides an upper bound for the duality gap. Thus, if we find a primal feasible point $\bar{x}$ and a corresponding dual feasible point

$(\bar{x}, \bar{\lambda})$ with $f(\bar{x}) - L(\bar{x}, \bar{\lambda}) = 0$, then $\bar{x}$ is a global minimal point of P and $(\bar{x}, \bar{\lambda})$ is a global maximal point of D.

Since the condition

$$0 = f(\bar{x}) - L(\bar{x}, \bar{\lambda}) = -\sum_{i \in I} \bar{\lambda}_i \, g_i(\bar{x}),$$

together with primal and dual feasibility, exactly means that $\bar{x}$ is a KKT point of P with multiplier vector $\bar{\lambda}$, we have thus provided an alternative proof of Theorem 3.2.78, based on duality theory. More importantly, next we will see that the termination criterion $t^k < \varepsilon$ from Algorithm 3.3 also provides an upper bound for the duality gap.

For this, we write down what Fermat's rule (Theorem 2.1.13) yields for the unconstrained minimal point x^k of $P(t^k)$ (using the chain rule to take the derivative of the logarithmic barrier function β):

$$0 = \nabla_x B(t^k, x^k) = \nabla f(x^k) + t^k \nabla \beta(x^k)$$

$$= \nabla f(x^k) + \sum_{i \in I} \left(-\frac{t^k}{g_i(x^k)} \right) \nabla g_i(x^k).$$

If we combine the values

$$\lambda_i^k := -\frac{t^k}{g_i(x^k)}, \quad i \in I,$$

into a vector λ^k, the latter satisfies the equation

$$\nabla_x L(x^k, \lambda^k) = 0.$$

From $t^k > 0$ and $g_i(x^k) < 0$, $i \in I$, we also obtain $\lambda^k > 0$, so that (x^k, λ^k) is a feasible point of the Wolfe-dual problem D. Since x^k is also primal feasible, these considerations yield

$$0 \leq v_P - v_D \leq f(x^k) - L(x^k, \lambda^k) = -\sum_{i \in I} \lambda_i^k \, g_i(x^k)$$

$$= -\sum_{i \in I} \left(-\frac{t^k}{g_i(x^k)} \right) g_i(x^k) = p \, t^k.$$

An interpretation of the termination criterion $t^k < \varepsilon$ in Algorithm 3.3 is therefore that the duality gap at termination drops below the value $p \, \varepsilon$.

Since the barrier method is a purely primal method, we formulate a purely primal result from these considerations for the termination criterion.

Theorem 3.3.15 *The application of Algorithm 3.3 with the logarithmic barrier function $\beta(x) = -\sum_{i\in I} \log(-g_i(x))$ to a convexly described C^1-problem P yields a point $\bar{x} \in M$ satisfying*

$$v_P \leq f(\bar{x}) < v_P + p\varepsilon.$$

Proof The primal feasibility of the last iterate $\bar{x} = x^k$ implies the first inequality. From $f(x^k) - L(x^k, \lambda^k) = pt^k < p\varepsilon$ and $L(x^k, \lambda^k) \leq v_D \leq v_P$ also $f(x^k) - v_P < p\varepsilon$ follows, thus the second inequality. $\qquad\square$

According to Theorem 3.3.15, $\bar{x}$ is a so-called $(p\varepsilon)$-minimal point of P.

From the above derivation, as a side result also an alternative motivation for the numerical instability of the barrier method for small barrier parameters t^k can be obtained. For this, we assume that the points x^k converge to a KKT point $x^\star$ of P, at which the LICQ holds with associated multiplier $\lambda^\star$. When using the logarithmic barrier function, as seen above, it must then hold

$$\lambda_i^k = -\frac{t^k}{g_i(x^k)} \rightarrow \lambda_i^\star.$$

The potential for numerical instability manifests itself in the fact that for all $i \in I_0(\bar{x})$ a quotient tends towards a real number, where the numerator and denominator simultaneously approach zero.

Exercise 3.3.16 Calculate the Hessian matrix $D_x^2 B(t^k, x^k)$ when using the logarithmic barrier function for $k \in \mathbb{N}$. What can you conclude about the curvature behavior of the functions $B(t^k, \cdot)$ at x^k for $k \to \infty$?

3.3.4 Primal-Dual Interior Point Methods

The first-order optimality condition derived in the previous section for x^k in the logarithmic barrier problem $P(t^k)$ is equivalent to the solvability of the system

$$\nabla f(x) + \sum_{i\in I} \lambda_i \nabla g_i(x) = 0,$$

$$g_i(x) < 0, \; i \in I,$$

$$\lambda_i > 0, \; i \in I,$$

$$\lambda_i \cdot g_i(x) + t = 0, \; i \in I$$

by (t^k, x^k, λ^k). This system is closely related to the KKT system of P, in which the complementarity conditions are *perturbed* by the parameter t. The crucial observation is that in this formulation no numerical instability is to be expected when t tends to zero. This leads to the following class of Karush-Kuhn-Tucker methods.

The basic idea of these methods is, while driving t to zero, to find zeros of

$$\mathscr{T}(t, x, \lambda) = \begin{pmatrix} \nabla f(x) + \sum_{i \in I} \lambda_i \nabla g_i(x) \\ \operatorname{diag}(\lambda) \cdot g(x) + t \cdot e \end{pmatrix}$$

with $g(x(t)) < 0$ and $\lambda(t) > 0$, and where e again denotes the all-ones vector. In the case of convergence $x(t) \to \bar{x}$ and $\lambda(t) \to \bar{\lambda}$, $\bar{x}$ is then a KKT point of P with multiplier $\bar{\lambda}$. The advantage of this approach is that neither problems with numerical instability nor with the domain of a barrier function occur. Indeed, one could search for zeros of $\mathscr{T}(t, x, \lambda)$ even for $t < 0$, even though one is not interested in these parameters.

> **Theorem 3.3.17** *Let $\bar{x}$ be a nondegenerate local minimal point of P with multiplier $\bar{\lambda}$. Then there are a neighborhood V of 0 and C^1-functions $x(t)$, $\lambda(t)$ on V with $(x(0), \lambda(0)) = (\bar{x}, \bar{\lambda})$, $\mathscr{T}(t, x(t), \lambda(t)) = 0$ for all $t \in V$, and $g(x(t)) < 0$ and $\lambda(t) > 0$ for all positive $t \in V$.*

Proof Due to $\mathscr{T}(0, \bar{x}, \bar{\lambda}) = 0$ and the nonsingularity of $D_{(x,\lambda)}\mathscr{T}(0, \bar{x}, \bar{\lambda})$, the implicit function theorem provides the existence of the functions $x(t)$ and $\lambda(t)$ on V with $(x(0), \lambda(0)) = (\bar{x}, \bar{\lambda})$ and $\mathscr{T}(t, x(t), \lambda(t)) = 0$ for all $t \in V$. According to the SCC, $\bar{\lambda}_i > 0$ holds for all $i \in I_0(\bar{x})$, so that for sufficiently small V also $\lambda_i(t) > 0$ is fulfilled on V. For all $i \in I_0(\bar{x})$, $\lambda_i(t)g_i(x(t)) + t = 0$ also implies $g_i(x(t)) < 0$ for positive t. For all $i \in I \setminus I_0(\bar{x})$, a continuity argument yields $g_i(x(t)) < 0$ and with an analogous argument $\lambda_i(t) > 0$ for $t > 0$.   □

> **Definition 3.3.18 (Primal-Dual Central Path)** With the notation from Theorem 3.3.17, the set
>
> $$C_{PD} = \{ (x(t), \lambda(t)) | t \in V, \ t > 0 \}$$
>
> is called *primal-dual central path* at $(\bar{x}, \bar{\lambda})$.

For linear and convex problems, the locally defined primal-dual central path can be extended globally and traced algorithmically. This results in efficient solution methods, which we cannot dive into in the context of this textbook [26]. However, we note that this derivation of the primal-dual interior-point methods crucially depends on the use of the *logarithmic* barrier function in the barrier problem. An alternative derivation motivates these methods by 'smoothing' the complicating complementarity conditions in the KKT system by perturbing them with t, thus making them algorithmically easier to handle. From this perspective, it is quite surprising that the perturbed KKT system again represents the first-order optimality condition of some optimization problem, namely the one of the associated barrier problem.

Clever implementations of primal-dual interior-point methods solve the perturbed KKT systems for large t only roughly, but become more precise for decreasing values of t. In addition, the methods determine the adjustment of t intrinsically, instead of iteratively multiplying t with a constant $\rho \in (0, 1)$ as in Algorithm 3.3. This can be done so efficiently in convexly described C^1-problems that the computational effort for identifying an ε-accurate minimal point, even in the worst case, only grows polynomially with the problem size.

Since this is particularly true for linear optimization problems, primal-dual interior-point methods are superior to the simplex algorithm in this respect, which needs exponential computational effort in the worst case. For high-dimensional problems, this superiority can actually be observed in practice, so that modern software packages usually do not solve large-scale linear optimization problems by the simplex algorithm, but with primal-dual interior-point methods. This is surprising insofar as, unlike in the simplex algorithm, the linearity of the defining functions of the optimization problem is not exploited at all, but only their convexity. Details on the design and convergence properties of primal-dual interior-point methods can be found, for example, in [11, 21, 27]. Some remarks on the transfer of primal-dual interior-point methods to certain classes of nonsmooth convex problems are given in [36].

3.3.5 Sequential Quadratic Programming Methods

Sequential quadratic programming (SQP) methods employ the idea of the Newton method to approximate a solution to a nonlinear problem by successively solving linearizations.

If one naively chooses the optimization problem P itself as the nonlinear problem, then the linearization is a linear optimization problem, which can be solved, for example, with the simplex algorithm or with primal-dual interior-point methods. The solution of this linearization is then used to linearize P again, solve the linearization, and so on. This approach leads, for example, to the *method of feasible directions* by Zoutendijk. Due to poor identification of active indices, this method can lead to so-called *jamming* and convergence to a noncritical point (for details see [7]). For a more successful development, known as *sequential linear programming (SLP)*, see [38].

However, in the first place the linearization of the problem P itself does not generalize the Newton idea from unconstrained optimization (Sect. 2.2.5). There, the objective function is *not* linearized (which, in the absence of constraints would usually lead to an unbounded problem), but the first-order optimality condition $\nabla f(x) = 0$ is: One sets $x^{k+1} = x^k + d^k$ with a solution d^k of the first-order optimality condition linearized around x^k,

$$\nabla f(x^k) + D^2 f(x^k) \cdot d^k = 0.$$

As we have already seen in Exercise 2.2.46, this equation is again a first-order optimality condition, namely for the quadratic optimization problem

$$Q^k : \quad \min_{d \in \mathbb{R}^n} \langle \nabla f(x^k), d \rangle + \tfrac{1}{2} d^\mathsf{T} D^2 f(x^k) d.$$

Upon convergence of the sequence (x^k) to a nondegenerate local minimal point, the matrix $D^2 f(x^k)$ is positive definite for all sufficiently large k for continuity reasons. Then Q^k is additionally a convex optimization problem, and the solution d^k of the linearized first-order condition is the global minimal point of Q^k.

The undamped Newton method for unconstrained problems thus essentially corresponds to solving a sequence of quadratic optimization problems, leading to the term sequential quadratic programming (SQP).

One may suspect that, for a generalization of the Newton method to the constrained case, the functions f, g_i, $i \in I$, h_j, $j \in J$, must be quadratically approximated and the resulting problems with quadratic objective function and quadratic constraints have to be solved. Instead, however, the auxiliary problems turn out to possess a much simpler structure.

To see this, we first consider the Newton method for solving the KKT system of P in the absence of inequality constraints, i.e., for $I = \emptyset$. We are thus looking for a root of

$$\mathcal{T}(x, \mu) = \begin{pmatrix} \nabla_x L(x, \mu) \\ h(x) \end{pmatrix}.$$

Given x^k and μ^k, the Newton method sets

$$\begin{pmatrix} x^{k+1} \\ \mu^{k+1} \end{pmatrix} = \begin{pmatrix} x^k \\ \mu^k \end{pmatrix} + \begin{pmatrix} d^k \\ \sigma^k \end{pmatrix}$$

with a solution (d^k, σ^k) of

$$0 = \mathcal{T}(x^k, \mu^k) + D\mathcal{T}(x^k, \mu^k) \begin{pmatrix} d \\ \sigma \end{pmatrix}. \tag{3.30}$$

Lemma 3.3.19 *Let $\bar{x}$ be a nondegenerate local minimal point of P with multiplier $\bar{\mu}$, and let the starting point (x^0, μ^0) lie sufficiently close to $(\bar{x}, \bar{\mu})$. Then (x^k, μ^k) converges quadratically to $(\bar{x}, \bar{\mu})$.*

Proof We have $\mathscr{T}(\bar{x}, \bar{\mu}) = 0$, and $D\mathscr{T}(\bar{x}, \bar{\mu})$ is nonsingular. Under our general smoothness assumption, we can assume f and h to be three times continuously differentiable, so that $D\mathscr{T}$ is locally Lipschitz continuous. From the convergence theory of the Newton method known from Sect. 2.2.5, the assertion follows. $\quad\square$

One may next ask whether, as in the unconstrained case, the linearized KKT system (3.30) 'coincidentally' is the KKT system of some auxiliary problem

$$Q^k: \quad \min_{d \in \mathbb{R}^n} F^k(d) \quad \text{s.t.} \quad H^k(d) = 0,$$

where d^k would be a KKT point of Q^k with multiplier σ^k. To verify this, we explicitly write out (3.30) and obtain

$$\begin{pmatrix} 0 \\ 0 \end{pmatrix} = \begin{pmatrix} \nabla_x L(x^k, \mu^k) \\ h(x^k) \end{pmatrix} + \begin{pmatrix} D_x^2 L(x^k, \mu^k) & \nabla h(x^k) \\ Dh(x^k) & 0 \end{pmatrix} \begin{pmatrix} d \\ \sigma \end{pmatrix},$$

which is equivalent to

$$\nabla_x L(x^k, \mu^k) + D_x^2 L(x^k, \mu^k)d + \nabla h(x^k)\sigma = 0,$$

$$h(x^k) + Dh(x^k)d = 0.$$

On the other hand, the KKT system of Q^k reads

$$\nabla F^k(d) + \nabla H^k(d)\sigma = 0,$$

$$H^k(d) = 0,$$

so that a comparison of the respective second equations leads to

$$H^k(d) = h(x^k) + Dh(x^k)d.$$

Thus, the summands $\nabla H^k(d)\sigma$ and $\nabla h(x^k)\sigma$ in the first equations also agree. It remains to construct a function F^k with

$$\nabla F^k(d) = \nabla_x L(x^k, \mu^k) + D_x^2 L(x^k, \mu^k)d,$$

for example

$$F^k(d) = \langle \nabla_x L(x^k, \mu^k), d \rangle + \tfrac{1}{2} d^\mathsf{T} D_x^2 L(x^k, \mu^k) d.$$

The first summand can be further simplified, because under the constraint $H^k(d) = 0$ we have

$$\langle \nabla_x L(x^k, \mu^k), d \rangle = \langle \nabla f(x^k), d \rangle + \mu^\mathsf{T} Dh(x^k) d = \langle \nabla f(x^k), d \rangle - \mu^\mathsf{T} h(x^k),$$

and $\mu^\mathsf{T} h(x^k)$ does not depend on d.

We have thus found that the solution (d^k, σ^k) of (3.30) may be interpreted as a KKT point and its multiplier of

$$Q^k: \quad \min_{d \in \mathbb{R}^n} \langle \nabla f(x^k), d \rangle + \tfrac{1}{2} d^\mathsf{T} D_x^2 L(x^k, \mu^k) d \quad \text{s.t.} \quad h(x^k) + Dh(x^k) d = 0.$$

So, as opposed to our above conjecture, only the objective function of the auxiliary problem is quadratic, while the constraints are even linear. Problems of this type are called *quadratic optimization problems*, so that we obtain again an *SQP method*. Observe that for $J = \emptyset$ we recover the problem Q^k from the unconstrained case, so we have indeed developed a generalization of the Newton method from the unconstrained to the constrained case.

Whether, as in the unconstrained case, Q^k is also a *convex* optimization problem when (x^k) converges to a nondegenerate local minimal point, is not immediately apparent. As in Example 3.2.64, at a nondegenerate local minimal point, the matrix $D_x^2 L(x^k, \mu^k)$ does not have to be positive definite on the whole space $\mathbb{R}^n$, but only on the tangent space to the feasible set. Therefore, the convexity of the objective function of Q^k on the whole space $\mathbb{R}^n$ cannot be guaranteed.

On the other hand, the objective function only needs to be convex on the feasible set of Q^k, and the latter is reminiscent of an approximation of the tangent space to M. If the Jacobian $Dh(x^k)$ possesses full rank, we can indeed argue as follows: For a vector $\bar{v}^k$ that solves the inhomogeneous linear system of equations $h(x^k) + Dh(x^k)d = 0$, and for an $(n, n-q)$-matrix V^k whose columns form a basis of the solution space of the homogeneous linear system of equations $Dh(x^k)d = 0$, it holds

$$\{ d \in \mathbb{R}^n \mid h(x^k) + Dh(x^k)d = 0 \} = \{ \bar{v}^k + V^k \eta \mid \eta \in \mathbb{R}^{n-q} \}.$$

The constrained problem Q^k can therefore be equivalently written as the unconstrained problem

$$\min_{\eta \in \mathbb{R}^{n-q}} \langle \nabla f(x^k), \bar{v}^k + V^k \eta \rangle + \tfrac{1}{2} \left(\bar{v}^k + V^k \eta \right)^\mathsf{T} D_x^2 L(x^k, \mu^k) \left(\bar{v}^k + V^k \eta \right).$$

In view of Definition 3.2.65 and continuity reasons, the Hessian matrix $(V^k)^\mathsf{T} D_x^2 L(x^k, \mu^k) V^k$ of the new quadratic objective function is positive definite for all sufficiently large k, when (x^k) converges to a nondegenerate local minimal point of P, so the unconstrained problem is convex.

Since, using linear algebra, the described reduction of the constrained problem Q^k to a lower-dimensional unconstrained problem can also be implemented algorithmically, we have simultaneously derived a method for solving the quadratic auxiliary problems Q^k, namely the *reduced SQP method*.

Next, let $I \neq \emptyset$ be allowed again. A general SQP method defines the new iterate as $x^{k+1} = x^k + t^k d^k$ by means of a KKT point d^k of

$$Q^k: \quad \min_{d \in \mathbb{R}^n} \langle \nabla f(x^k), d \rangle + \tfrac{1}{2} d^\mathsf{T} L^k d \quad \text{s.t.} \quad g(x^k) + Dg(x^k)d \leq 0,$$

$$h(x^k) + Dh(x^k)d = 0,$$

where L^k stands for $D_x^2 L(x^k, \lambda^k, \mu^k)$ or an approximation of this matrix (which leads, e.g., to quasi-Newton SQP methods) and where t^k is determined by a step size control (Sect. 3.3.6).

Due to the polyhedrality of the feasible set of Q^k and Corollary 3.2.24, every minimal point d^k of Q^k is automatically a KKT point. If, in addition, L^k is positive semidefinite, then Q^k is also a convex optimization problem, so that according to Theorem 3.2.78 every KKT point d^k is also a global minimal point of Q^k. As above in the case $I = \emptyset$, it would be sufficient to impose such (semi-)definiteness assumptions on L^k only on the solution space of the homogeneous system $Dg(x^k)d \leq 0$, $Dh(x^k)d = 0$, which we will refrain from in this textbook for the sake of clarity.

If Q^k possesses the vector $d^k = 0$ as a KKT point, then the SQP method can terminate with a KKT point x^k of P.

Lemma 3.3.20 *For $d^k = 0$, x^k is a KKT point of P.*

Proof Since d^k is a KKT point of Q^k, there exist $\tau^k \geq 0$ and $\sigma^k \in \mathbb{R}^q$ with

$$\nabla f(x^k) + L^k d^k + \nabla g(x^k)\tau^k + \nabla h(x^k)\sigma^k = 0, \tag{3.31}$$

$$(\tau^k)^\mathsf{T} \left(g(x^k) + Dg(x^k)d^k \right) = 0, \tag{3.32}$$

$$h(x^k) + Dh(x^k)d^k = 0, \tag{3.33}$$

$$g(x^k) + Dg(x^k)d^k \leq 0. \tag{3.34}$$

For $d^k = 0$ this yields the claim. $\qquad\square$

Algorithmically, Lemma 3.3.20 is employed by terminating already for $\|d^k\|_2 \le \varepsilon_1$ with some user-defined tolerance $\varepsilon_1 > 0$. The point x^k is then an approximation of a KKT point of P with multipliers τ^k and σ^k as in the proof of Lemma 3.3.20, where from (3.31) the estimate

$$\|\nabla f(x^k) + \nabla g(x^k)\tau^k + \nabla h(x^k)\sigma^k\|_2 \;=\; \|L^k d^k\|_2 \;\le\; \|L^k\|_2\,\varepsilon_1$$

follows with the spectral norm $\|L^k\|_2$ of L^k. To also obtain estimates for the approximate validity of the remaining conditions in the KKT system, the termination criterion may be augmented with a ε_2-feasibility condition $\max\{\,|Dh_j(x^k)d^k|,\ j \in J,\ -Dg_i(x^k)d^k,\ i \in I\} \le \varepsilon_2$, where $\varepsilon_2 > 0$ denotes a feasibility tolerance parameter. A method for solving quadratic optimization problems like Q^k is discussed in detail in Sect. 3.3.7.

3.3.6 Merit Functions and Filter Methods

Next we turn to the step size control in the SQP method, i.e., the determination of some $t^k > 0$, with which after the calculation of d^k, τ^k and σ^k

$$\begin{pmatrix} x^{k+1} \\ \lambda^{k+1} \\ \mu^{k+1} \end{pmatrix} \;=\; \begin{pmatrix} x^k \\ \lambda^k \\ \mu^k \end{pmatrix} + t^k \begin{pmatrix} d^k \\ \tau^k \\ \sigma^k \end{pmatrix}$$

is set. As in the unconstrained case, one may first ask whether d^k is a descent direction for f at x^k, so that t^k could then be chosen such that a sufficiently large descent in f is realized by the new iterate. According to Lemma 3.3.20, we may assume $d^k \ne 0$ as soon as an iteration step is required.

If the iterate x^k is feasible for P, i.e., if $g(x^k) \le 0$ and $h(x^k)=0$ hold, then sometimes d^k indeed provides a first-order descent direction for f. In fact, due to the feasibility of x^k for P, also $d=0$ is feasible for Q^k, and its minimal value is therefore bounded above by zero. If L^k is also positive definite, we obtain

$$\langle \nabla f(x^k), d^k \rangle \;\le\; -\tfrac{1}{2}(d^k)^\mathsf{T} L^k d^k \;<\; 0,$$

so that d^k is a first-order descent direction for f at x^k.

Unfortunately, x^k is *not* necessarily feasible for P. Then not only the above argument is impossible, but indeed the direction d^k must also ensure 'increasing feasibility' of the next iterate and *cannot* always be a descent direction for f.

We will see below, however, that d^k is at least a descent direction for the auxiliary objective function of the ℓ_1-penalty method

$$A_1(\rho, x) \;=\; f(x) + \rho\,\alpha_1(x)$$

with sufficiently large $\rho > 0$ is (Sect. 3.3.1). This means that, for sufficiently small $t^k > 0$, the new iterate $x^{k+1} = x^k + t^k d^k$ at least reduces a weighted sum of the objective function f and the feasibility measure α_1.

Merit Functions

Every such function, for which it is meaningful to desire a descent by a step in direction d^k, is called a *merit function* for P. The following result states that, for example, $A_1(\rho, \cdot) > 0$ is a merit function for all sufficiently large ρ. According to Exercise 3.3.3, $A_1(\rho, \cdot)$ is one-sided directionally differentiable at every $x \in \mathbb{R}^n$ with computable directional derivative $A_1'(\rho, x, d)$, so that we may consider every direction $d \in \mathbb{R}^n$ with $A_1'(\rho, x, d) < 0$ as a first-order descent direction for $A_1(\rho, \cdot)$ in x.

Theorem 3.3.21 *Let* $L^k \succ 0$, *and let* $d^k \neq 0$ *be a KKT point of* Q^k *with multipliers* τ^k, σ^k. *Then* d^k *is a first-order descent direction for* $A_1(\rho, x) = f(x) + \rho \alpha_1(x)$ *in* x^k *whenever* $\rho \geq \|(\tau^k, \sigma^k)\|_\infty$ *holds.*

Proof Let the index $k \in \mathbb{N}$ be fixed and omitted, i.e., we set $x = x^k$, $d = d^k$ etc. For each $\rho > 0$, according to Exercise 3.3.3, $A_1(\rho, x)$ is one-sided directionally differentiable at x in direction d with

$$A_1'(\rho, x, d) = \langle \nabla f(x), d \rangle + \rho \left(\sum_{i \in I_0(x)} \max\{0, \langle \nabla g_i(x), d \rangle\} + \sum_{i \in I_+(x)} \langle \nabla g_i(x), d \rangle \right.$$

$$\left. - \sum_{j \in J_-(x)} \langle \nabla h_j(x), d \rangle + \sum_{j \in J_0(x)} |\langle \nabla h_j(x), d \rangle| + \sum_{j \in J_+(x)} \langle \nabla h_j(x), d \rangle \right).$$

We will use in the following that d satisfies the KKT conditions (3.31) to (3.34). From (3.31) it follows

$$A_1'(\rho, x, d) = \underbrace{-d^\mathsf{T} L d}_{< 0} - \sum_{i \in I} \tau_i \langle \nabla g_i(x), d \rangle - \sum_{j \in J} \sigma_j \langle \nabla h_j(x), d \rangle$$

$$+ \rho \left(\sum_{i \in I_0(x)} \max\{0, \langle \nabla g_i(x), d \rangle\} + \sum_{i \in I_+(x)} \langle \nabla g_i(x), d \rangle \right.$$

$$\left. - \sum_{j \in J_-(x)} \langle \nabla h_j(x), d \rangle + \sum_{j \in J_0(x)} |\langle \nabla h_j(x), d \rangle| + \sum_{j \in J_+(x)} \langle \nabla h_j(x), d \rangle \right).$$

With (3.32) to (3.34) we conclude

$$A_1'(\rho, x, d) < \sum_{i \in I} \tau_i g_i(x) + \sum_{j \in J} \sigma_j h_j(x)$$

$$+ \rho \left(\sum_{i \in I_0(x)} \max\{0, -g_i(x)\} - \sum_{i \in I_+(x)} g_i(x) + \sum_{j \in J_-(x)} h_j(x) - \sum_{j \in J_+(x)} h_j(x) \right)$$

$$= \underbrace{\sum_{i \in I_-(x)} \tau_i g_i(x)}_{\leq 0} + \sum_{i \in I_+(x)} \tau_i g_i(x) + \sum_{j \in J_-(x)} \sigma_j h_j(x) + \sum_{j \in J_+(x)} \sigma_j h_j(x)$$

$$+ \rho \left(- \sum_{i \in I_+(x)} g_i(x) + \sum_{j \in J_-(x)} h_j(x) - \sum_{j \in J_+(x)} h_j(x) \right)$$

$$\leq \sum_{i \in I_+(x)} (\tau_i - \rho) g_i(x) + \sum_{j \in J_-(x)} (\sigma_j + \rho) h_j(x) + \sum_{j \in J_+(x)} (\sigma_j - \rho) h_j(x) \leq 0,$$

where the last inequality follows from the definitions of $I_+(x)$, $J_-(x)$, $J_+(x)$ and $\rho \geq \|(\tau^k, \sigma^k)\|_\infty = \max\{\tau_i, i \in I, |\sigma_j|, j \in J\}$.   □

According to Theorem 3.3.21, for sufficiently large ρ the function $A_1(\rho, x)$ is not only a merit function, but it is even one-sided directionally differentiable with $A_1'(\rho, x^k, d^k) < 0$ for $d^k \neq 0$. Therefore, the descent in $A_1(\rho, x)$ from x^k in direction d^k can actually be implemented algorithmically by some inexact line search, for example by the Armijo rule with the adjustment for one-sided directionally differentiable functions mentioned at the end of Sect. 2.2.2. Moreover, here the required size of ρ is *known*, because according to Theorem 3.3.21 it is determined by the known multipliers τ^k and σ^k to the minimal point d^k of Q^k.

In summary, the presented idea for step size control in SQP methods is to achieve a descent in the merit function $A_1(\rho, x)$ for an iterate x^k and a solution d^k of the SQP subproblem Q^k by the Armijo rule, where a suitable value for ρ is calculated from the multipliers associated with d^k. With regard to the following idea, recall that the Armijo rule reduces too large and thus unacceptable values of t by backtracking line search (i.e., by multiplication with a reduction factor) until the value of the merit function is suitably reduced by the new iterate.

Filter Methods

The above step size control by inexact one-dimensional minimization of the merit function $f(x) + \rho \, \alpha_1(x)$ is ultimately based on the weighted sum scalarization for the multicriteria problem

$$\min_x \begin{pmatrix} f(x) \\ \alpha_1(x) \end{pmatrix}$$

discussed in Sect. 3.3.1. *Filter methods* handle this multicriteria problem in a different way, which in particular does not require the choice of a parameter ρ. They use the concept of *dominated points* of multicriteria problems.

Definition 3.3.22 (Filter)

(a) For two iterates x^k and x^ℓ the pair $(f(x^k), \alpha_1(x^k))$ *dominates* the pair $(f(x^\ell), \alpha_1(x^\ell))$, if both $f(x^k) \le f(x^\ell)$ and $\alpha_1(x^k) \le \alpha_1(x^\ell)$ hold, and if at least one of these two inequalities is strict.
(b) A list of pairs $(f(x^k), \alpha_1(x^k))$, none of which dominates another, is called a *filter*.
(c) An iterate x^{k+1} is called *acceptable*, if the pair $(f(x^{k+1}), \alpha_1(x^{k+1}))$ is not dominated by any filter entry.

The basic procedure of a *filter SQP method* consists in including a pair $(f(x^{k+1}), \alpha_1(x^{k+1}))$ in the filter if the iterate x^{k+1} is acceptable, followed by deletion of all filter entries dominated by this new entry. Figure 3.21 illustrates a filter with four entries and the level line of a merit function $f + \rho\alpha_1$ in the (f, α_1)-space through the point $(f(x^k), \alpha_1(x^k))$. While the point x^{k+1} is accepted by the filter, but not by the merit function, the situation is reversed for the point $\tilde{x}^{k+1}$. This illustrates that filters generally accept other new iterates than merit functions.

That a new iterate $x^k + td^k$ is not accepted by the filter is the analogue of the merit function case in which the step size t is too large to reduce the value of the merit function. As described, the basic idea of the Armijo rule is then to reduce t by backtracking line search until the value of the merit function is sufficiently reduced by the new iterate. Similarly, also the filter idea can be combined with backtracking line search, where t is multiplied by a reduction factor until the new iterate is accepted by the filter. This procedure is referred to as a *backtracking line search filter SQP method*.

If $x^k + td^k$ is not accepted by the filter, an alternative way to generate a new iterate closer to x^k is provided by a trust-region approach. Then the problem Q^k is modified by the additional constraint $\|d\|_\infty \le t^k$, and the trust-region radius t^k is adapted with the ideas from Sect. 2.2.10 until the new iterate is accepted by the

Fig. 3.21 Filter and merit function

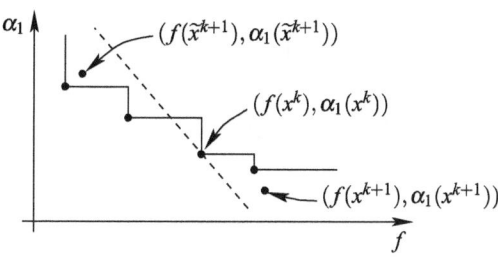

filter. The resulting class of *trust-region filter SQP methods* provides an effective and stable algorithmic solution strategy for many constrained nonlinear optimization problems. For details on the convergence of these approaches, we refer to [26].

3.3.7 Quadratic Optimization

Quadratic optimization problems often occur in nonlinear optimization (and beyond) as auxiliary problems, such as the problem Q^k in the SQP method or the trust-region subproblem TR^k with a polyhedral constraint like $\|d\|_\infty \le t^k$. Also the orthogonal projection $\mathrm{pr}(x, M)$ of a point x onto a polyhedral set M is the minimal point of a quadratic optimization problem.

Quadratic optimization problems have the general form

$$Q: \quad \min_{x \in \mathbb{R}^n} \tfrac{1}{2}x^\mathsf{T} L x + c^\mathsf{T} x \quad \text{s.t.} \quad a_i^\mathsf{T} x + b_i \le 0, \quad i \in I,$$

$$a_j^\mathsf{T} x + b_j = 0, \quad j \in J,$$

where in the cases of interest to us even $L = L^\mathsf{T} \succeq 0$ holds, so that Q is a convex problem. To avoid confusion of indices, in this section we define $I = \{1, \dots, p\}$ and $J = \{p + 1, \dots, p + q\}$, and we will derive a method specifically tailored to this structure of Q.

We first consider the case $I = \emptyset$ and set

$$A := \begin{pmatrix} a_1^\mathsf{T} \\ \vdots \\ a_q^\mathsf{T} \end{pmatrix} \quad \text{and} \quad b := \begin{pmatrix} b_1 \\ \vdots \\ b_q \end{pmatrix}.$$

We may assume that $\mathrm{rank}(A) = q$ holds, because otherwise the feasible set of Q is either empty, or one can eliminate redundant equality constraints until the rank condition applies. Then a first possibility to solve Q is to parametrize the solution space of $Ax + b = 0$ and to solve the corresponding unconstrained quadratic problem. This explicit and possibly cumbersome parametrization can be avoided by a second approach.

Indeed, according to Corollary 3.2.24 and Theorem 3.2.78, $\bar{x}$ is a minimal point of Q if and only if it is a KKT point of Q, i.e., there exists some $\bar{\mu}$ with

$$\begin{pmatrix} 0 \\ 0 \end{pmatrix} = \begin{pmatrix} L\bar{x} + c + A^\mathsf{T}\bar{\mu} \\ A\bar{x} + b \end{pmatrix} = \begin{pmatrix} L & A^\mathsf{T} \\ A & 0 \end{pmatrix} \begin{pmatrix} \bar{x} \\ \bar{\mu} \end{pmatrix} + \begin{pmatrix} c \\ b \end{pmatrix}. \tag{3.35}$$

The linear system of equations (3.35) can be solved, for example, by Gaussian elimination.

Apparently, Q is solvable if and only if (3.35) is solvable. In the following we are interested in unique solvability. According to Lemma 3.2.61, the matrix

$$\begin{pmatrix} L & A^\mathsf{T} \\ A & 0 \end{pmatrix}$$

is nonsingular if and only if $\operatorname{rank}(A^\mathsf{T}) = q$ holds and if $L|_{\ker A}$ is nonsingular. We have already assumed the first of these two conditions. If L is not only positive semi-definite, but even positive definite, the second condition is automatic. To see this, let $L \succ 0$ and V be a matrix whose columns form a basis of $\ker A$. Then for all $\eta \in \mathbb{R}^{n-q} \setminus \{0\}$

$$\eta^\mathsf{T} V^\mathsf{T} L V \eta = (V\eta)^\mathsf{T} L(V\eta) > 0$$

and thus $L|_{\ker A} \succ 0$ hold, which implies the regularity of this matrix. For $L \succ 0$, Q and (3.35) are therefore uniquely solvable.

In the case $L \succeq 0$, Q and (3.35) may be unsolvable. Using theorems of the alternative, it can be shown that Q is then necessarily unbounded (for details see [7]).

Next, we again allow $I \neq \emptyset$. The idea of *active set methods* is to consider the active inequality constraints as equality constraints in a feasible iteration point x^k, and to solve this equality-constrained problem with the above method. Since its solution $\bar{x}^k$ may be infeasible, one moves along the line segment from x^k to $\bar{x}^k$ until hitting the boundary of M, to find the new iterate x^{k+1}, as illustrated in Fig. 3.22 for x^0 and x^1. Since the point x^2 is a KKT point, the procedure can terminate.

If one starts in Fig. 3.22 with y^0, the procedure described so far gets stuck at the point y^1, although it is not a KKT point. There, however, at least one negative multiplier λ_j exists, and one can delete the corresponding constraint j to obtain a new auxiliary problem.

Fig. 3.22 Active set method

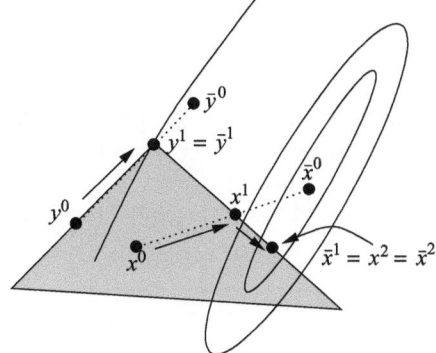

To describe the active set method in more detail, we consider the quadratic optimization problem

$$Q: \quad \min_{x \in \mathbb{R}^n} q(x) \quad \text{s.t.} \quad a_i^\mathsf{T} x + b_i \leq 0, \quad i \in I,$$

$$a_j^\mathsf{T} x + b_j = 0, \quad j \in J,$$

with objective function

$$q(x) = \tfrac{1}{2} x^\mathsf{T} L x + c^\mathsf{T} x$$

and $L \succ 0$. The KKT system of Q at a feasible point x^k with $I_0^k := I_0(x^k)$ is

$$\nabla q(x^k) + \sum_{i \in I_0^k} \lambda_i \, a_i + \sum_{j \in J} \mu_j \, a_j = 0, \tag{3.36}$$

$$a_i^\mathsf{T} x^k + b_i = 0, \quad i \in I_0^k \cup J, \tag{3.37}$$

$$a_i^\mathsf{T} x^k + b_i < 0, \quad i \notin I_0^k, \tag{3.38}$$

$$\lambda_i \geq 0, \quad i \in I_0^k. \tag{3.39}$$

If one ignores the sign conditions (3.38) and (3.39), then the remaining Eqs. (3.36) and (3.37) form the KKT system of

$$Q^k: \quad \min_{x \in \mathbb{R}^n} q(x) \quad \text{s.t.} \quad a_i^\mathsf{T} x + b_i = 0, \, i \in I_0^k \cup J.$$

In the following, we assume that the matrix formed by the rows a_i^T, $i \in I_0^k \cup J$, has full rank for all k. The unique minimal point $\bar{x}^k$ of Q^k can then be determined, for example, by Gaussian elimination for the system (3.35) associated to Q^k.

For the convergence proof of the active set method, the behavior of the minimal values $q(x^k)$ will be crucial. In this regard, we note that due to the feasibility of x^k for Q^k, the inequality $q(\bar{x}^k) \leq q(x^k)$ is fulfilled.

Case 1: $x^k \neq \bar{x}^k$. Since Q^k is uniquely solvable under our assumptions, in this case even $q(\bar{x}^k) < q(x^k)$ holds. To cover the case of a point $\bar{x}^k$ that is infeasible for Q, we define $x^{k+1} = x^k + t^k(\bar{x}^k - x^k)$ with the maximal step size $t^k \in (0, 1]$ such that x^{k+1} is still feasible for Q. This step size can be calculated as

$$t^k = \min \left(1, \min_{\substack{i \in I \\ a_i^\mathsf{T}(\bar{x}^k - x^k) > 0}} -\frac{a_i^\mathsf{T} x^k + b_i}{a_i^\mathsf{T}(\bar{x}^k - x^k)} \right).$$

The convexity of q yields

$$q(x^{k+1}) \leq (1 - t^k)q(x^k) + t^k q(\bar{x}^k) < q(x^k).$$

For $t^k = 1$ we have $x^{k+1} = \bar{x}^k$ and, thus, usually $I_0^{k+1} = I_0^k$, but in exceptional cases also $I_0^{k+1} \supseteq I_0^k$. In the following iteration step, $x^{k+1} = \bar{x}^{k+1}$ will therefore hold, i.e., the second case treated below occurs.

For $t^k < 1$, at least one new constraint $i_+ \notin I_0^k$ becomes active, so $I_0^{k+1} \supseteq I_0^k \cup \{i_+\}$. Due to the rank condition in Q^k, this addition of an active constraint can occur at most $(n - q)$ times in succession.

Case 2: $x^k = \bar{x}^k$. First observe that, according to the above considerations, this case occurs at least once during $n - q$ iterations.

Due to the rank condition, x^k solves the system (3.36) and (3.37) with unique multipliers $\lambda_i, i \in I_0^k, \mu_j, j \in J$. If $\lambda_i \geq 0$ is fulfilled for all $i \in I_0^k$, (3.38) also holds, and x^k is a KKT point of Q. The procedure can terminate in this case. Otherwise, there exists some index $i_- \in I_0^k$ with $\lambda_{i_-} < 0$. We remove the constraint with the index i_- and consider the problem

$$\widetilde{Q}^k : \quad \min_{x \in \mathbb{R}^n} q(x) \quad \text{s.t.} \quad a_i^\mathsf{T} x + b_i = 0, \ i \in (I_0^k \setminus \{i_-\}) \cup J.$$

Let $\widetilde{x}^k$ be the unique minimal point of this problem. The point x^k is feasible for $\widetilde{Q}^k$, but cannot be optimal for this problem. Since the multipliers are uniquely determined, the KKT condition (3.36) with $\lambda_{i_-} = 0$ would otherwise have to be fulfilled, which contradicts $\lambda_{i_-} < 0$. It follows $q(\widetilde{x}^k) < q(x^k)$.

Next, we have to consider a possible infeasibility of $\widetilde{x}^k$ for Q as in the first case. Initially it is not clear whether along the direction $(\widetilde{x}^k - x^k)$ feasible points may be found at all. However, by Theorem 2.1.40 the convexity of q implies

$$0 > q(\widetilde{x}^k) - q(x^k) \geq \langle \nabla q(x^k), \widetilde{x}^k - x^k \rangle$$

$$\overset{(3.36)}{=} -\sum_{i \in I_0^k} \lambda_i a_i^\mathsf{T}(\widetilde{x}^k - x^k) - \sum_{j \in J} \mu_j a_j^\mathsf{T}(\widetilde{x}^k - x^k) = \underbrace{-\lambda_{i_-} \cdot (a_{i_-}^\mathsf{T} \widetilde{x}^k + b_{i_-})}_{> 0},$$

so that i_- cannot lie in $I_0(\widetilde{x}^k)$. In particular, $(\widetilde{x}^k - x^k)$ is a feasible direction in x^k, and we set $x^{k+1} = x^k + t^k(\widetilde{x}^k - x^k)$ with t^k as above. From the convexity of q it then follows again $q(x^{k+1}) < q(x^k)$.

Theorem 3.3.23 *Let $L \succ 0$, and for all k let the system of equations in Q^k possess full rank. Then the active set method finds the minimal point of Q in a finite number of steps.*

Proof The second case discussed above occurs at least once during $n - q$ iterations. In the second case, x^k minimizes the function q over the facet

$$M^k = \{x \in M \mid a_i^\mathsf{T} x + b_i = 0, \; i \in I_0^k \cup J\}$$

of the polyhedral feasible set M of Q. Furthermore, the sequence $(q(x^k))$ is strictly decreasing, so that during the iteration in the second case no facet of M can be visited twice. Since M only possesses a finite number of facets, the assertion follows.   □

In the case $L = 0$, the active set method collapses to the (primal) simplex algorithm of linear optimization. Also a feasible starting point x^0 for the active set method can be found as in phase 1 of the simplex algorithm [25].

For alternative algorithmic approaches to quadratic optimization which are based on, for example, ideas of interior-point methods or projected gradients, we refer to [26].

Bibliography

1. Alt, W.: Nichtlineare Optimierung. Vieweg, Wiesbaden (2002)
2. Bazaraa, M.S., Sherali, H.D., Shetty, C.M.: Nonlinear Programming. Wiley, New York (1993)
3. Beck, A.: Introduction to Nonlinear Optimization. MOS-SIAM Series on Optimization, Philadelphia (2014)
4. Bonnans, J.F., Shapiro, A.: Perturbation Analysis of Optimization Problems. Springer, New York (2000)
5. Broyden, C.G.: The convergence of a class of double-rank minimization algorithms. IMA J. Appl. Math. **6**, 76–90 (1970)
6. Davidon, W.C.: Variable metric method for minimization. SIAM J. Optim. **1**, 1–17 (1991)
7. Faigle, U., Kern, W., Still, G.: Algorithmic Principles of Mathematical Programming. Kluwer, New York (2002)
8. Fischer, G.: Lineare Algebra. SpringerSpektrum, Berlin (2014)
9. Fletcher, R.: A new approach to variable metric algorithms. Comput. J. **13**, 317–322 (1970)
10. Fletcher, R., Powell, M.J.D.: A rapidly convergent descent method for minimization. Comput. J. **6**, 163–168 (1963)
11. Freund, R.W., Hoppe, R.H.W.: Stoer/Bulirsch: Numerische Mathematik 1. Springer, Berlin (2007)
12. Geiger, C., Kanzow, C.: Numerische Verfahren zur Lösung unrestringierter Optimierungsaufgaben. Springer, Berlin (1999)
13. Geiger, C., Kanzow, C.: Theorie und Numerik restringierter Optimierungsaufgaben. Springer, Berlin (2002)
14. Goldfarb, D.: A family of variable metric updates derived by variational means. Math. Comput. **24**, 23–26 (1970)
15. Gould, F.J., Tolle, J.W.: A necessary and sufficient qualification for constrained optimization. SIAM J. Appl. Math. **20**, 164–172 (1971)
16. Güler, O.: Foundations of Optimization. Springer, Berlin (2010)
17. Heuser, H.: Lehrbuch der Analysis, Teil 2. SpringerVieweg, Wiesbaden (2008)
18. Heuser, H.: Lehrbuch der Analysis, Teil 1. SpringerVieweg, Wiesbaden (2009)
19. Jahn, J.: Introduction to the Theory of Nonlinear Optimization. Springer, Berlin (1994)
20. Jänich, K.: Lineare Algebra. Springer, Berlin (2008)
21. Jarre, J., Stoer, J.: Optimierung. Springer, Berlin (2004)
22. Jongen, H.Th., Jonker, P., Twilt, F.: Nonlinear Optimization in Finite Dimensions. Kluwer, Dordrecht (2000)
23. Jongen, H.Th., Meer, K., Triesch, E.: Optimization Theory. Kluwer, Dordrecht (2004)
24. Lasserre, J.B.: On representations of the feasible set in convex optimization. Optim. Lett. **4**, 1–5 (2010)
25. Nickel, S., Rebennack, S., Stein, O., Waldmann, K.-H.: Operations Research, 3rd edn. SpringerGabler, Berlin (2022)

26. Nocedal, J., Wright, S.: Numerical Optimization. Springer, New York (2006)
27. Reemtsen, R.: Lineare Optimierung. Shaker, Maastricht (2001)
28. Rockafellar, R.T., Wets, R.J.B.: Variational Analysis. Springer, Berlin (1998)
29. Rudin, W.: Principles of Mathematical Analysis. McGraw-Hill, New York (1976)
30. Shanno, D.F.: Conditioning of quasi-Newton methods for function minimization. Math. Comput. **24**, 647–656 (1970)
31. Stein, O.: Bi-level Strategies in Semi-infinite Programming. Kluwer, Boston (2003)
32. Stein, O.: On constraint qualifications in non-smooth optimization. J. Optim. Theory Appl. **121**, 647–671 (2004)
33. Stein, O.: Grundzüge der Konvexen Analysis. SpringerSpektrum, Berlin (2021)
34. Stein, O.: Grundzüge der Parametrischen Optimierung. SpringerSpektrum, Berlin (2021)
35. Stein, O.: Grundzüge der Gemischt-ganzzahligen Optimierung. SpringerSpektrum, Berlin (2024)
36. Stein, O.: Basic Concepts of Global Optimization. Springer, Berlin (2024)
37. Werner, J.: Numerische Mathematik II. Vieweg-Verlag, Braunschweig (1992)
38. Zhang, J.: Superlinear convergence of a trust-region-type successive linear programming method. J. Optim. Theory Appl. **61**, 295–310 (1989)

Index